I0469010

Regional Mitigation Strategy for the Arizona Solar Energy Zones

Final Report

prepared by
Environmental Science Division
Argonne National Laboratory

for
U.S. Department of the Interior
Bureau of Land Management

March 2016

This page intentionally left blank

TABLE OF CONTENTS

NOTATION vii

ABSTRACT 1

1 INTRODUCTION AND PURPOSE 3

    1.1 Purpose of the Strategy 3
    1.2 Background 4
    1.3 Solar Regional Mitigation Strategy Development Process 8
    1.4 Stakeholder Engagement & Involvement in the Solar Regional Mitigation Strategy 10

2 REGIONAL COMPENSATORY MITIGATION STRATEGY – ARIZONA SOLAR ENERGY ZONES 13

    2.1 Description of the Arizona Solar Energy Zones and Surrounding Region 13
        2.1.1 General Description of the Agua Caliente Solar Energy Zone 13
        2.1.2 General Description of the Brenda Solar Energy Zone 13
        2.1.3 General Description of the Gillespie Solar Energy Zone 13
        2.1.4 Landscape Conditions of the Arizona Solar Energy Zones and the Region 17
        2.1.5 Regional Setting 17
            2.1.5.1 General Description 17
            2.1.5.2 Regional Conditions and Trends 24
    2.2 General Description of Solar Development in the Arizona Solar Energy Zones 27
        2.2.1 Description of Existing Rights-of-Way, Development Status, and Recommended Non-Development Areas 27
            2.2.1.1 Agua Caliente Solar Energy Zone 27
            2.2.1.2 Brenda Solar Energy Zone 27
            2.2.1.3 Gillespie Solar Energy Zone 29
        2.2.2 Description of Potential Development 30
    2.3 Summary of Solar Development Impacts on the Arizona Solar Energy Zones 32
    2.4 Mitigation Strategy (Hierarchy) for the Arizona Solar Energy Zones 32
        2.4.1 Avoidance 32
        2.4.2 Minimization 33
            2.4.2.1 Summary of Programmatic Design Features to be Applied 33
            2.4.2.2 Other Required Impact Minimization Measures and/or Stipulations 33
        2.4.3 Regional Compensatory Mitigation 34
            2.4.3.1 Identification of Residual Impacts 35
            2.4.3.2 Residual Impacts that May Warrant Regional Compensatory Mitigation 36
                2.4.3.2.1 Conceptual Models 36
                2.4.3.2.2 Residual Impacts Warranting Compensatory Mitigation 36
    2.5 Regional Goals and Mitigation Desired Outcomes 38
    2.6 Calculating the Recommended Mitigation Obligation for Arizona Solar Energy Zones 44
    2.7 Management of Solar Regional Compensatory Mitigation Obligations 49
    2.8 Evaluation of Compensatory Mitigation Sites, Actions, and Desired Outcomes 49
    2.9 Mitigation Effectiveness Monitoring and Adaptive Management Plan 53
    2.10 Implementation Strategy 60

3 REFERENCES 61

TABLE OF CONTENTS (Cont.)

4   GLOSSARY 63

APPENDIX A Impact Assessment Summary Tables A-1

APPENDIX B Conceptual Models B-1

APPENDIX C Summary Tables: Impacts that May Warrant Regional Mitigation for the Three
Arizona Solar Energy Zones Agua Caliente, Brenda and Gillespie C-1

APPENDIX D BLM Screening of Candidate Regional Mitigation Sites for the Arizona Solar Energy
Zones D-1

FIGURES

1-1   Mitigation flow diagram for solar energy development 6

1-2   Timeline of solar regional compensatory mitigation processes relative to solar energy development schedule 9

2-1   Agua Caliente Solar Energy Zone and surrounding areas as identified in the RDEP ROD 14

2-2   Brenda Solar Energy Zone and surrounding areas as identified in the Solar PEIS ROD 15

2-3   Gillespie Solar Energy Zone and surrounding areas as identified in the Solar PEIS ROD 16

2-4   Current Sonoran Desert terrestrial landscape intactness in six classes from high to very low depicted with a 4 km X 4 km grid cell 18

2-5   Arizona Solar Energy Zones overlain on terrestrial intactness model 19

2-6   Land cover types in the vicinity of the Brenda Solar Energy Zone, Gillespie Solar Energy Zone, and Agua Caliente Solar Energy Zone 21

2-7   Land cover types of the Sonoran Desert ecoregion in Arizona 22

2-8   Agua Caliente Solar Energy Zone recommended developable area 28

2-9   Brenda Solar Energy Zone recommended developable area 28

2-10   Gillespie Solar Energy Zone recommended developable area 29

2-11   Steps for calculating per-acre regional compensatory mitigation fees based on impacts 45

2-12   Highest scoring candidate site locations based on the Candidate Site Matrix, Appendix D 52

D-1   Recommended Candidate Regional Mitigation Sites for Arizona SEZs, western sites D-4

D-2   Recommended Candidate Regional Mitigation Sites for Arizona SEZs, eastern sites D-5

TABLES

1-1   Fees and costs associated with renewable energy development 7

2-1   Land cover types and amounts in the vicinity of the Agua Caliente Solar Energy Zone 20

2-2   Land cover types and amounts in the vicinity of the Brenda Solar Energy Zone 23

2-3   Land cover types and amounts in the vicinity of the Gillespie Solar Energy Zone 24

2-4   Ecological intactness of the Arizona Solar Energy Zones and condition assessment for ecological systems 26

2-5   Summary table of regional goals, objectives, and mitigation desired outcomes and actions for the Arizona Solar Energy Zones 39

2-6   Components of the recommended per acre compensatory mitigation fees for the Arizona Solar Energy Zones 48

2-7   Recommended methods and measurements for core and contingent indicators 56

2-8   Quantitative indicators and measurements relevant to each of the three land health attributes 57

A-1   Agua Caliente Solar Energy Zone: Impact Assessment Summary Table A-3

A-2   Brenda Solar Energy Zone: Impact Assessment Summary Table A-21

A-3   Gillespie Solar Energy Zone: Impact Assessment Summary Table A-39

C-1   Agua Caliente Solar Energy Zone – Summary Table: Impacts That May Warrant Regional Compensatory Mitigation C-3

C-2   Brenda Solar Energy Zone – Summary Table: Impacts That May Warrant Regional Compensatory Mitigation C-10

C-3   Gillespie Solar Energy Zone – Summary Table: Impacts That May Warrant Regional Compensatory Mitigation C-16

D-1   BLM Matrix for Candidate Regional Mitigation Sites for Arizona SEZs D-7

NOTATION

ACRONYMS, INITIALISMS, AND ABBREVIATIONS

AAQS	Ambient Air Quality Standards
ACEC	Area of Critical Environmental Concern
AIM	Assessment, Inventory, and Monitoring
AZGFD	Arizona Game and Fish Department
BLM	Bureau of Land Management
DOE	Department of Energy
DOI	Department of the Interior
EIS	Environmental Impact Statement
EO	Executive Order
ESA	Endangered Species Act
FEMA	Federal Emergency Management Agency
FO	Field Office
IDT	Interdisciplinary Team
LCRMSCP	Lower Colorado River Multi-Species Conservation Program
LPI	Line Point Intercept
LWC	Lands with Wilderness Characteristics
MQ	Management Question
MTR	Military Training Route
MW	Megawatt
NEPA	National Environmental Policy Act
NFWF	National Fish & Wildlife Foundation
NHT	National Historic Trail
NWR	National Wildlife Refuge
PEIS	Programmatic Environmental Impact Statement
PM	Particulate Matter
PV	Photovoltaic
RDEP	Restoration Design Energy Project
REA	Rapid Ecoregional Assessment
REDA	Renewable Energy Development Areas
RMP	Resource Management Plan
ROD	Record of Decision

ROW	Right-of-Way
SDA	Specially Designated Areas
SEZ	Solar Energy Zone
SHPO	State Historic Preservation Office(r)
SRMA	Special Recreation Management Area
SRMS	Solar Regional Mitigation Strategy
SSS	Special Status Species
TNC	The Nature Conservancy
TWS	The Wilderness Society
USFWS	United States Fish & Wildlife Service
USGS	United States Geological Service
VRI	Visual Resource Inventory
VRM	Visual Resource Management
WA	Wilderness Area
WSA	Wilderness Study Area

UNITS OF MEASURE

cm	centimeter
km^2	square kilometers(s)
m	meter(s)

UNIT CONVERSIONS

$1\ km^2$	$0.39\ mi^2$
1 m	3.28 ft

ABSTRACT

The "Regional Mitigation Strategy for the Arizona Solar Energy Zones" presents a strategy for compensating for the residual or unavoidable impacts that are expected from the development of the Agua Caliente, Brenda, and Gillespie Solar Energy Zones (SEZs) in western Arizona. This strategy responds to a call for the development of solar regional compensatory mitigation strategies for each of the SEZs, as committed to in the Record of Decision for the Final Programmatic Environmental Impact Statement (PEIS) for Solar Energy Development in Six Southwestern States. The strategy consists of preliminary findings and recommendations for identifying: (1) the residual impacts of utility-scale solar development in the Arizona SEZs that may warrant regional compensatory mitigation, (2) mitigation actions that can be implemented in the region to compensate for those impacts, (3) how appropriate regional compensatory mitigation obligations could be determined, and (4) how the impacts and mitigation actions could be monitored. While this strategy for the Arizona SEZs is not a Bureau of Land Management decision, it will inform future decisions, including project-specific National Environmental Protection Act (NEPA), regarding configuration of lease parcels and issuance of leases for the Arizona SEZs, lease stipulations, impacts warranting compensatory mitigation in the region, where and how regional compensatory mitigation might occur, and how monitoring and adaptive management might occur.

This page intentionally left blank

1 INTRODUCTION AND PURPOSE

1.1 Purpose of the Strategy

This "Regional Mitigation Strategy for the Arizona Solar Energy Zones" recommends compensatory mitigation for certain residual impacts expected from the development of Arizona Solar Energy Zones (SEZs) in the western part of the state. As mandated by the Federal Land Policy and Management Act of 1976, the Bureau of Land Management (BLM) is required to manage public lands for multiple uses while protecting the quality of ecological and other environmental and cultural values, in a manner that does not result in the permanent impairment of the productivity of the land. While the BLM places a priority on avoiding and minimizing impacts, especially onsite, avoidance and minimization[1] may not be sufficient. Utility-scale solar development often involves a long-term commitment of resources over a relatively large area. The BLM is considering requirements for compensatory mitigation for those residual impacts that warrant regional compensatory mitigation. Accordingly, this strategy provides:

1. The residual impacts expected as a result of development within the Agua Caliente, Brenda, and Gillespie SEZs (Appendix A).

2. The regionally important trends in the Sonoran Desert ecoregion where the Arizona SEZs are located (Section 2.1.5.2).

3. Conceptual models that depict the relationships between resources, ecosystem functions and services, and change agents (i.e., human development and use, climate change, wildfire, and invasive species) (Section 2.4.3.2.1; Appendix B).

4. The residual impacts that, in consideration of regional trends and the roles that the impacted resources play, may warrant regional compensatory mitigation (Section 2.4.3.2.2; Appendix C).

5. Regional goals and objectives for resources identified with residual impacts warranting mitigation, including those recommended in the applicable land use plan(s), and mitigation desired outcomes (Section 2.5).

6. A recommended method for calculating a regional compensatory mitigation fee that could be assessed to developers choosing to contribute to a mitigation fund, and an explanation of how it was calculated for each of the Arizona SEZs. Also, the strategy includes the estimated cost of regional compensatory mitigation action(s) that would compensate for residual impacts and help meet regional goals and objectives, including a breakout of acquisition, restoration, and/or ongoing management costs to ensure effectiveness, additionality, and durability (Section 2.6).

7. Preliminary information on management of mitigation obligation revenues derived from development of the Arizona SEZs (Section 2.7).

8. Recommended regional compensatory mitigation sites, action(s), and desired outcomes for the Arizona SEZs to contribute to achieving the regional goals and objectives (Section 2.8).

---

[1] Terms used throughout this document are defined in the Glossary.

9. Discussion of how the mitigation outcomes should be monitored and what will happen if the actions are not achieving the desired results (Section 2.9).

The BLM authorized officer will make a determination of compensatory mitigation requirements prior to issuing the lease and notice to proceed and will also take into consideration:

- The National Environmental Policy Act (NEPA) analyses completed for the lease sale, project permitting, and mitigation alternatives, including opportunity for public and stakeholder participation and comments.

- Any changes to the applicable resource management plans (RMPs) or other related plans that affect management of the SEZs or possible mitigation sites.

- The input received from Government-to-Government consultation with tribes.

- Any other information that would update, correct, or otherwise supplement the information contained in this strategy.

1.2 Background

In 2012, the BLM and the U.S. Department of Energy (DOE) published the "Final Programmatic Environmental Impact Statement (PEIS) for Solar Energy Development in Six Southwestern States" (Final Solar PEIS; BLM and DOE 2012). The Final Solar PEIS assessed the impact of utility-scale solar energy development on public lands in the six southwestern states of Arizona, California, Colorado, Nevada, New Mexico, and Utah. The "Approved Resource Management Plan Amendments/Record of Decision (ROD) for Solar Energy Development in Six Southwestern States" (Solar PEIS ROD) implemented a comprehensive solar energy program for public lands in those states (also called the Western Solar Plan) and incorporated land use allocations and programmatic and SEZ-specific design features into land use plans in the six-state study area (BLM 2012a). Seventeen priority areas for utility-scale solar energy development, or SEZs, were identified in the Solar PEIS ROD, including the Brenda and Gillespie SEZs in Arizona. The Final Solar PEIS presents a detailed analysis of the expected impacts of solar development on each SEZ.

BLM Arizona has conducted a statewide planning effort for facilitating renewable energy development (the Restoration Design Energy Project or RDEP; ROD signed in January 2013) that builds from the Solar PEIS. The Agua Caliente SEZ was established through the signing of the RDEP ROD (BLM 2013a), following the requirements of the BLM's solar energy program. The Draft and Final Environmental Impact Statement (EIS) for the RDEP project included detailed analyses and maps for the Agua Caliente SEZ supporting establishment of this SEZ, along with other Renewable Energy Development Areas (REDAs). Although this Solar Regional Mitigation Strategy (SRMS) document and its recommendations apply directly to the Agua Caliente, Brenda, and Gillespie SEZs, the information herein and the process outlined could be used and applied to REDAs or other lands in the future.

Comments on both the Draft Solar PEIS and the Supplement to the Draft Solar PEIS encouraged the BLM to incorporate a robust mitigation framework into the proposed solar energy program to address any residual impacts expected to result from solar development in the SEZs, despite avoidance of many impacts and the implementation of design features to minimize impacts. In the Supplement to the Draft Solar PEIS, the BLM presented, as part of its incentives for SEZs, the concept of regional mitigation planning[2]. A draft framework for regional mitigation planning was posted on the project web page between the publication of the Supplement to the Draft Solar PEIS and the Final Solar PEIS to foster stakeholder engagement. A revised framework for regional mitigation planning was then included in the Final Solar PEIS and the Solar PEIS ROD. The BLM is continuing to refine a process for developing solar regional mitigation strategies for SEZs, and has released a draft procedural guidance document on the topic (BLM 2014a).

Federal regulations require consideration of a mitigation hierarchy consisting of avoidance, minimization, rectification, reduction or elimination of impacts over time, and/or compensation (i.e., the mitigation hierarchy[3]) (40 CFR 1508.20). Implementation of the mitigation hierarchy begins with the location and configuration of the SEZs, so as to avoid as many conflicts as possible. Avoidance is also used within the boundaries of SEZs by designating non-development areas. Minimization involves the implementation of design features (which, in the case of the Solar PEIS, are required mitigation measures) and management practices meant to reduce the impacts on site. The Solar PEIS and the RDEP EIS analyzed the impacts of solar development assuming a robust suite of design features would be in place. The RODs for these documents adopted a robust set of both programmatic and SEZ-specific design features into the BLM's solar energy program in order to minimize some of the expected impacts of development onsite. These design features will be included as part of the Plans of Development required for projects within SEZs prior to BLM issuance of leases, or as stipulations in the leases. This SRMS addresses only the last aspect of the mitigation hierarchy, compensatory mitigation. Compensatory mitigation is evaluated by the BLM based on the need to address residual impacts to resources (i.e., those impacts that cannot be avoided or minimized; also referred to as "unavoidable impacts").

Figure 1-1 illustrates how mitigation measures identified in the Solar PEIS ROD and the Arizona RDEP EIS ROD, including design features, are carried forward and are included, to the extent they apply, in project-specific NEPA conducted following a submission of an application by a developer. It is important to note that avoidance of resource impacts was included in designating the SEZs and REDAs. Table 1-1 illustrates the context of the per acre mitigation fee recommended in this SRMS document in comparison to other fees and costs to be borne by the project developer through time. The fees and costs include rental and nameplate capacity fees, costs for implementing design features to accomplish on-site mitigation, compensatory mitigation fees, and bonding costs for reclamation of the project site following decommissioning.

This strategy consists of recommendations to compensate for some of the residual impacts that will remain after avoidance and minimization measures are taken. A major focus of this regional compensatory mitigation strategy is to provide a recommended fee to be paid by the developer that will offset those residual impacts and to offer a suite of mitigation actions and locations, depending on project-specific details, to meet mitigation goals and objectives for effectiveness, feasibility, durability, and additionality. This strategy differs from project-level compensatory mitigation development that has been conducted historically by the BLM because this regional strategy has been developed in

---

[2] In the Final Solar Energy PEIS (BLM and DOE 2012), Appendix A, Section A.2.5, the BLM refers to solar regional mitigation plans (SRMPs). To be consistent with guidance issued in the subsequent BLM Instruction Memorandum 2013-142 (BLM 2013b), the BLM herein adopts the terminology of solar regional mitigation *strategies* (SRMSs).

[3] Throughout this document, the terminology of avoidance and minimization may be used to also refer to other parts of the mitigation hierarchy, specifically rectification and reduction or elimination of impacts over time.

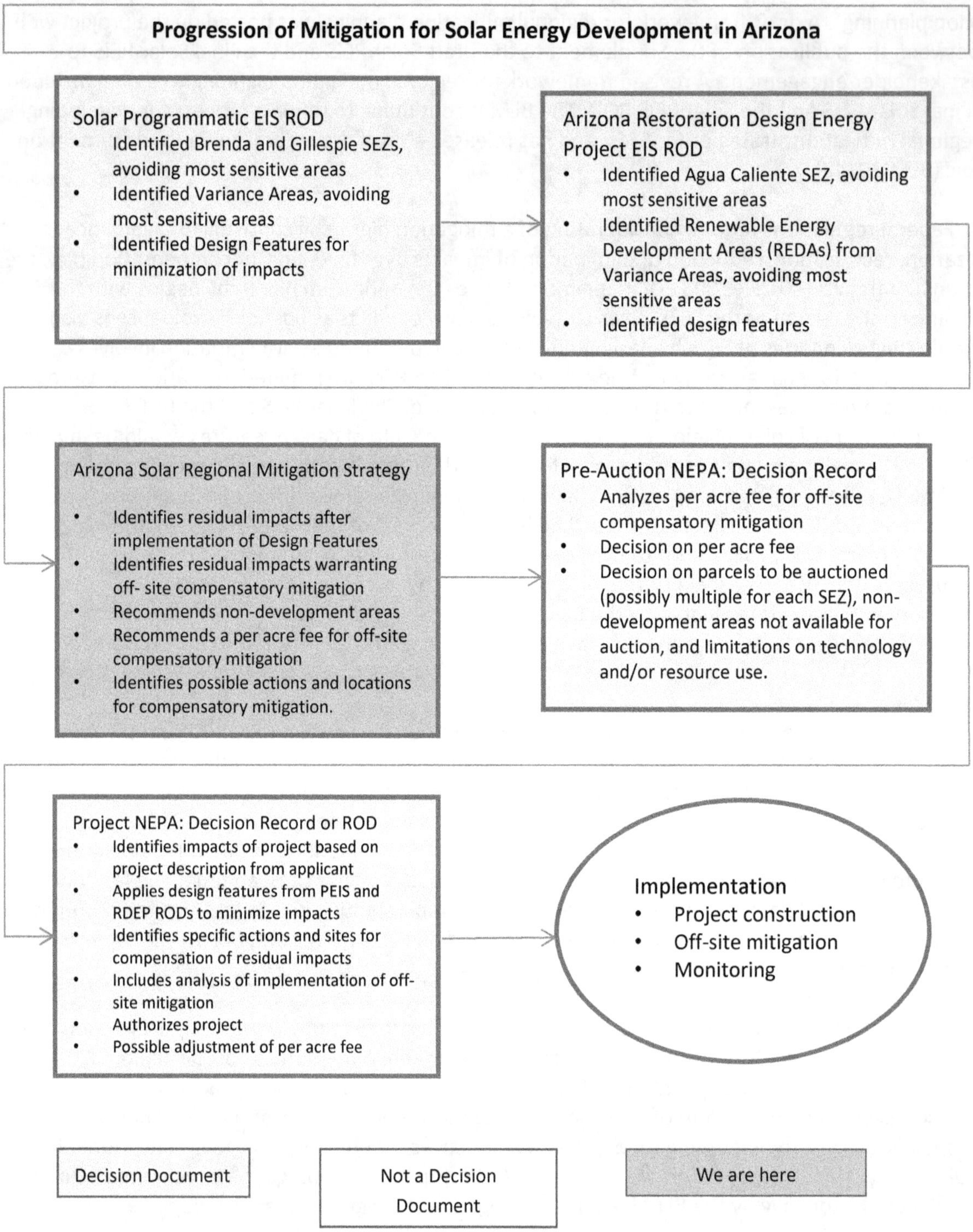

Progression of Mitigation for Solar Energy Development in Arizona

Solar Programmatic EIS ROD
- Identified Brenda and Gillespie SEZs, avoiding most sensitive areas
- Identified Variance Areas, avoiding most sensitive areas
- Identified Design Features for minimization of impacts

Arizona Restoration Design Energy Project EIS ROD
- Identified Agua Caliente SEZ, avoiding most sensitive areas
- Identified Renewable Energy Development Areas (REDAs) from Variance Areas, avoiding most sensitive areas
- Identified design features

Arizona Solar Regional Mitigation Strategy
- Identifies residual impacts after implementation of Design Features
- Identifies residual impacts warranting off- site compensatory mitigation
- Recommends non-development areas
- Recommends a per acre fee for off-site compensatory mitigation
- Identifies possible actions and locations for compensatory mitigation.

Pre-Auction NEPA: Decision Record
- Analyzes per acre fee for off-site compensatory mitigation
- Decision on per acre fee
- Decision on parcels to be auctioned (possibly multiple for each SEZ), non-development areas not available for auction, and limitations on technology and/or resource use.

Project NEPA: Decision Record or ROD
- Identifies impacts of project based on project description from applicant
- Applies design features from PEIS and RDEP RODs to minimize impacts
- Identifies specific actions and sites for compensation of residual impacts
- Includes analysis of implementation of off-site mitigation
- Authorizes project
- Possible adjustment of per acre fee

Implementation
- Project construction
- Off-site mitigation
- Monitoring

| Decision Document | Not a Decision Document | We are here |

Figure 1-1. Mitigation flow diagram for solar energy development

Table 1-1. Fees and costs associated with renewable energy development (green highlighted element addressed in this Solar Regional Mitigation Strategy)

Fee/Cost Borne By Developer	When Paid	Disposition
Accepted Bid at Auction	At Issuance of Lease	U.S. Treasury (BLM recovers reasonable costs)
Rent (per acre)	At Issuance of Lease	U.S. Treasury
Nameplate Capacity Fee (per megawatt)		
Per acre Mitigation Fee (Recommended in this SRMS)	At Issuance of Notice to Proceed	Held by BLM in a specific account or with third party, e.g., NFWF
Cost of implementation of design features and other project-specific mitigation	During project construction and operation	Spent by developer on project implementation activities
Bond for post-closure reclamation of project site	At Issuance of Lease	Held by BLM, returned if not needed by BLM
Reclamation of project site after decommissioning	Cost borne by Lease-holder, or BLM uses reclamation bond	Spent by developer (or BLM) on reclamation activities

advance of project-specific analyses, considers conditions and trends of various resources in the context of the larger landscape, and identifies the desired outcome for compensatory mitigation actions, including the outline for a comprehensive protocol for monitoring those actions. This SRMS is consistent with BLM's interim policy on regional mitigation, Draft Manual Section 1794, issued on June 13, 2013 (BLM 2013b).

1.3 Solar Regional Mitigation Strategy Development Process

In August 2012, the BLM initiated a regional mitigation strategy for solar energy development with the Dry Lake SEZ (Nevada), which constituted the first SRMS developed for an SEZ. The Dry Lake SEZ SRMS originated simultaneously with, and served as a pilot test case for, the establishment of BLM's interim policy on regional mitigation (Draft Manual Section 1794). The Dry Lake SEZ SRMS was completed in 2014 (BLM 2014b) and, together with the BLM's interim policy on regional mitigation[4], served as a guide for preparing this SRMS for the Arizona SEZs.

The process for developing the SRMS for the Arizona SEZs largely followed the outline for regional mitigation planning presented in the Final Solar PEIS. In general, a team of specialists from the BLM in Arizona, with the support of Argonne National Laboratory, produced a preliminary product at each step in the process, which was then presented and discussed in a public forum. The opportunity for written comments was also extended to the public. The methods used and content of this SRMS incorporate many of the ideas and comments received from the public.

The mitigation actions identified in this strategy are designed to compensate for residual impacts on habitat, cultural resources, visual resources, and ecological services that are expected from the development of the Agua Caliente, Brenda, and Gillespie SEZs. For the purpose of this analysis, it is assumed that all of the developable land within the three SEZs will be impacted. Recommendations on the degree of compensation considered the condition of the resource values present in the SEZs and also the relevant management objectives in the RMPs and the relative costs and benefits of the use of public lands for solar energy development, including the amount of time and effort required to restore the disturbed areas upon expiration of the leases. The recommended mitigation actions are drawn from stakeholder recommendations and from the Yuma RMP, the Lake Havasu RMP, and the Lower Sonoran RMP (BLM 2010, 2007, 2012b, respectively). These documents describe resource management goals and objectives and identify restoration and preservation needs within the landscape in which the SEZs are located.

Under the terms of this strategy, the amount of the recommended mitigation fees for the Arizona SEZs is based on the impacts of solar development in the SEZs. As part of the BLM solar energy program, long-term monitoring will be used to evaluate the effectiveness of the regional compensatory mitigation strategy for the Arizona SEZs (consistent with the BLM's Assessment, Inventory, and Monitoring [AIM] Strategy (Toevs et al. 2011)). This regional compensatory mitigation strategy will be subject to continued review and adjustment to ensure mitigation desired outcomes are being met.

The timeline of this SRMS process, relative to a solar development project implementation schedule, is provided in Figure 1-2. The compensatory mitigation obligation (fee) will be analyzed and established along with the environmental impacts of leasing parcels within the SEZ for future solar energy development during a pre-auction NEPA analysis. The compensatory mitigation obligation,

---

[4] Departmental Manual Part 600 DM 6 on Landscape-Scale Mitigation Policy was finalized on October 23, 2015 (DOI 2015).

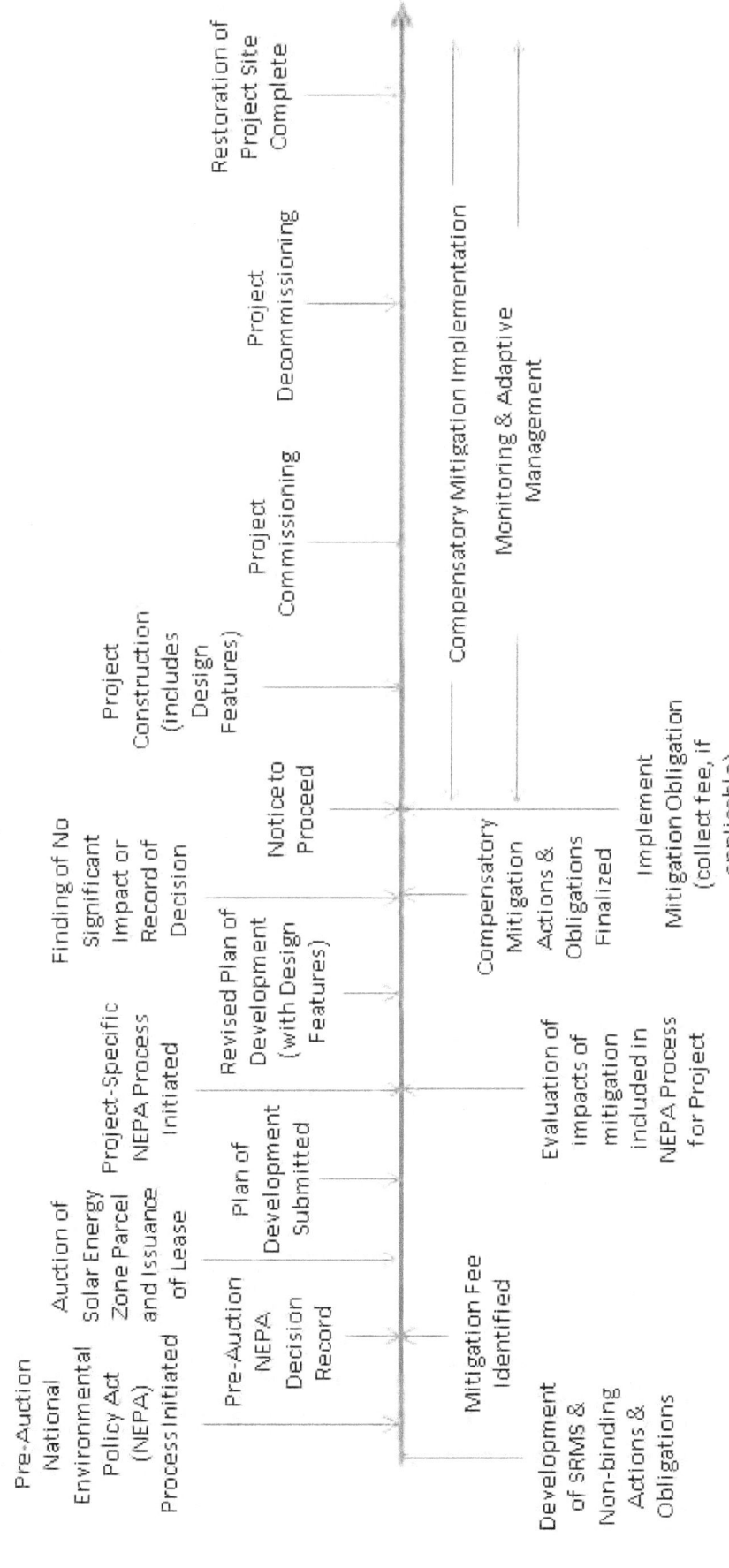

Figure 1-2. Timeline of solar regional compensatory mitigation processes relative to solar energy development schedule

site(s), and action(s) strategically recommended in this document will be considered in the project-specific NEPA evaluation required for planned solar energy developments within the Arizona SEZs (see Figure 1-1). At the conclusion of the project-specific NEPA evaluation, the BLM authorized officer will identify the appropriate compensatory mitigation obligation, site(s), and action(s) as part of the BLM's project decision. The compensatory mitigation obligation, site(s), and action(s) selected by the authorized officer may differ from the recommendations made in this SRMS document and may be based on several factors, including but not limited to (1) new information regarding the presence/absence of environmental resources that may change the potential for impact; (2) implementation of additional design features, avoidance areas, or other technologies not evaluated in the BLM Solar PEIS that would minimize impacts; (3) new information about additional mitigation sites or actions; and/or (4) updated assessments of mitigation costs and an adjustment of the base fee for inflation to current year dollars.

1.4 Stakeholder Engagement & Involvement in the Solar Regional Mitigation Strategy

Stakeholder engagement and involvement in developing the Arizona SRMS included three workshops in Phoenix and three web-based meetings. Representatives from federal, state, and local government agencies; non-governmental organizations concerned with issues such as environmental or recreational impacts; representatives from the solar development industry and utilities; tribal representatives; and individual members of the public were invited to attend these activities. Approximately 40 individuals and representatives from the previously mentioned organizations attended the kickoff workshop held April 2, 2014. During the first workshop, background on solar regional mitigation strategies and the Solar PEIS impact assessment for the Arizona SEZs were provided to the attendees.

The second workshop was held November 12–13, 2014. This workshop included a discussion of mitigation goals and objectives and a variety of stakeholders presented their recommendations for candidate regional compensatory mitigation sites and actions to be evaluated in the mitigation strategy. The second workshop had about 35 attendees, including individuals and representatives from agencies, nongovernmental organizations, the solar industry and consultants to the industry, utilities, and tribes.

The final workshop was held September 17, 2015. This workshop presented the draft strategy to the public and included an open forum for discussion, questions, comments, and clarifications. The third workshop included about 20 participants, including individuals and representatives from agencies, nongovernmental organizations, local universities, consultants to the solar industry, and tribes.

Additionally, the following webinars were held: on July 9, 2014, to provide information on revisions to SEZ residual impacts and impacts that may warrant regional compensatory mitigation and to request recommendations for candidate regional compensatory mitigation sites and actions; on February 24, 2015, to revisit mitigation goals and objectives and candidate mitigation sites and provide BLM recommendations for non-development areas within the SEZs; and on March 25, 2015 to discuss mitigation obligations.

All presentations from the workshops and webinars are posted on the project documents web page on the Arizona SEZs SRMS Project website at: http://www.blm.gov/az/st/en/prog/energy/solar/arizona_regional_mitigation.html. Additional materials that were provided for stakeholder review are posted on the project documents web page as well.

Throughout the project, stakeholders were invited to comment on interim draft materials, including the summary of residual impacts at the three Arizona SEZs that may warrant regional compensatory mitigation, the matrix used to evaluate candidate compensatory mitigation sites and activities proposed for the Arizona SEZs, and the recommended compensatory mitigation fee. Many of these comments were discussed during the workshops and webinars and were used to guide development of this strategy.

This page intentionally left blank

2 REGIONAL COMPENSATORY MITIGATION STRATEGY – ARIZONA SOLAR ENERGY ZONES

2.1 Description of the Arizona Solar Energy Zones and Surrounding Region

2.1.1 General Description of the Agua Caliente Solar Energy Zone

The Agua Caliente SEZ is located in Yuma County in southwestern Arizona. The total area of the Agua Caliente SEZ, as shown in Figure 2-1, is 2,560 acres (10.3 km^2). About 20,600 acres (83 km^2) were originally identified for analysis in the RDEP EIS, but the BLM revised the proposed SEZ boundaries in the Final EIS and included a smaller proposed SEZ in Alternative 6, the alternative that was ultimately selected in the ROD. The boundaries in the Final EIS were revised to exclude major washes, maintain an area for potential tortoise migration between the Palomas Mountains and Baragan Mountain, and avoid most known archaeological sites and lands with wilderness characteristics (BLM 2013a). Of the 2,560 acres (10.3 km^2) of developable area in the SEZ identified in the RDEP ROD, the SRMS recommends development on up to 2,021 acres (8.18 km^2) (see Section 2.4.1).

The cities of Yuma and Buckeye are located 65 mi (105 km) southwest and 60 mi (97 km) northeast of the SEZ, respectively. Dateland (population of 852 in 2000) is the nearest community and is located along the nearest major road to the Agua Caliente SEZ, Interstate 8, approximately 12 mi (19 km) south of the SEZ. Palomas Road is a Yuma County road that passes just south of the SEZ and provides direct access to the SEZ. The nearest railroad stop is 0.6 mi (1 km) away. The area around the SEZ is sparsely populated with limited economic development opportunities.

2.1.2 General Description of the Brenda Solar Energy Zone

The Brenda SEZ is located in La Paz County in west-central Arizona, 32 mi (52 km) east of the California border. The total area of the Brenda SEZ, as shown in Figure 2-2, is 3,348 acres (13.5 km^2) (BLM and DOE 2012). In the Final Solar PEIS, the boundaries were reduced to eliminate the Bouse Wash and an area on the west side of the SEZ. Of the 3,348 acres (13.5 km^2) of developable area in the SEZ identified in the Solar PEIS ROD, the SRMS recommends development on up to 1,906 acres (7.71 km^2) (see Section 2.4.1).

The town of Brenda is located about 3 mi (5 km) southwest of the SEZ. The towns of Quartzsite and Salome in La Paz County are about 18 mi (29 km) west of, and 18 mi (29 km) east of, the SEZ respectively. The Phoenix metropolitan area is approximately 100 mi (161 km) to the east of the SEZ. The nearest major road access is U.S. 60, which runs along the southeast border of the SEZ. The nearest railroad stop is 11 mi (18 km) away.

2.1.3 General Description of the Gillespie Solar Energy Zone

The Gillespie SEZ is located in Maricopa County in west-central Arizona. The total area of the Gillespie SEZ, as shown in Figure 2-3, is 2,618 acres (11 km^2) (BLM and DOE 2012). The SRMS recommends development on up to 2,231 acres (9.03 km^2) (see Section 2.4.1).

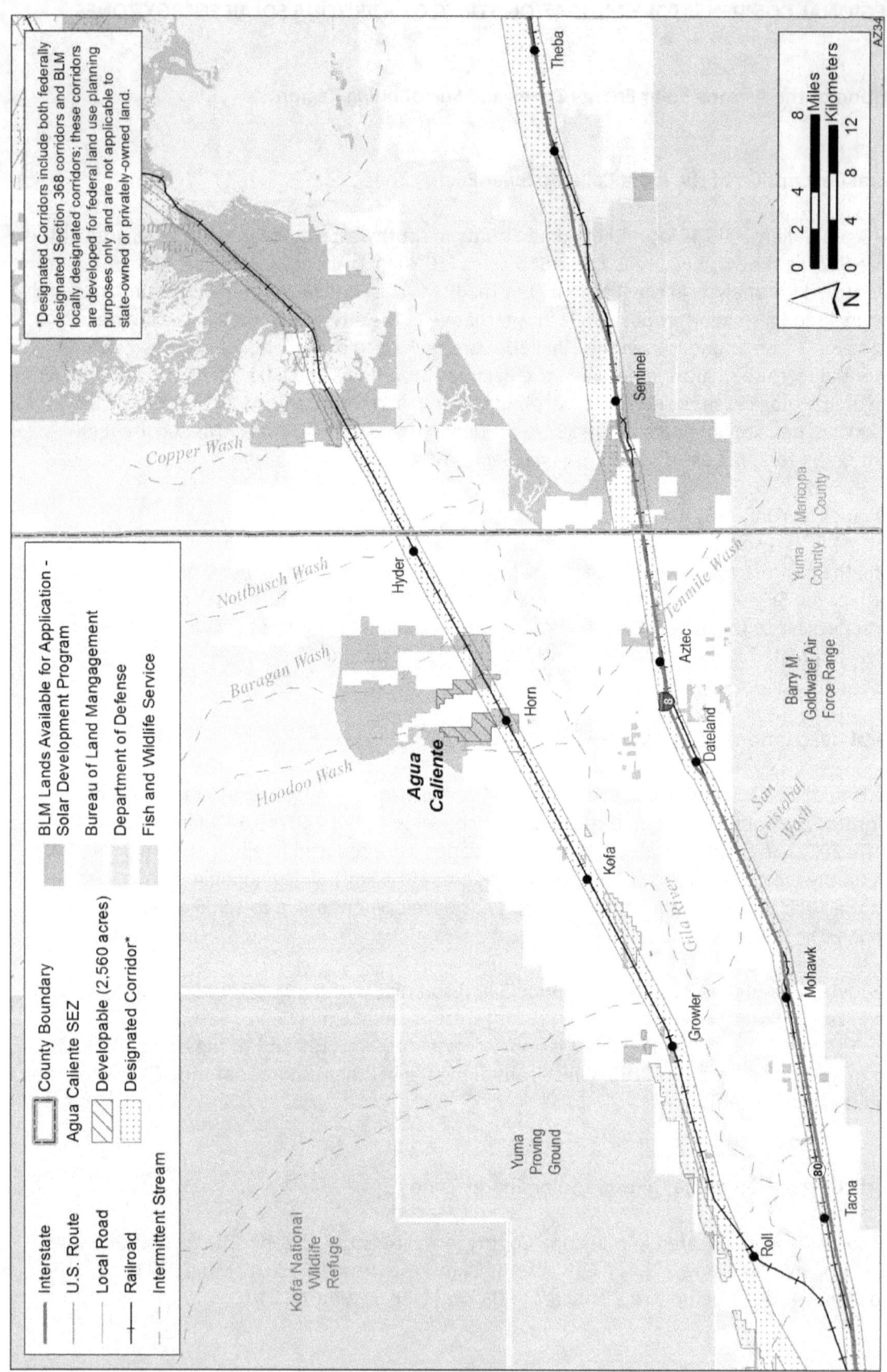

Figure 2-1. Agua Caliente Solar Energy Zone and surrounding areas as identified in the RDEP ROD

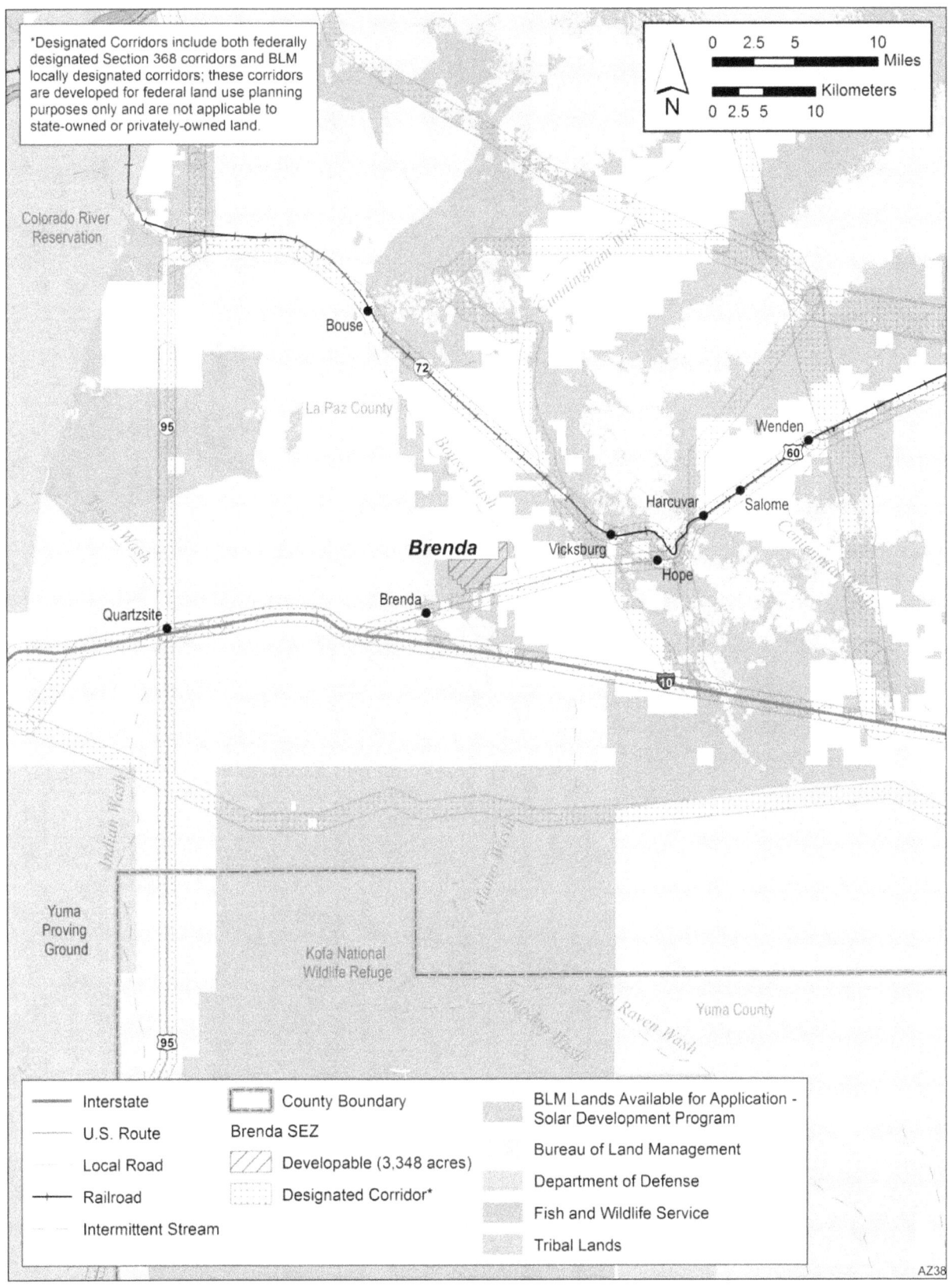

*Designated Corridors include both federally designated Section 368 corridors and BLM locally designated corridors; these corridors are developed for federal land use planning purposes only and are not applicable to state-owned or privately-owned land.

Figure 2-2. Brenda Solar Energy Zone and surrounding areas as identified in the Solar PEIS ROD

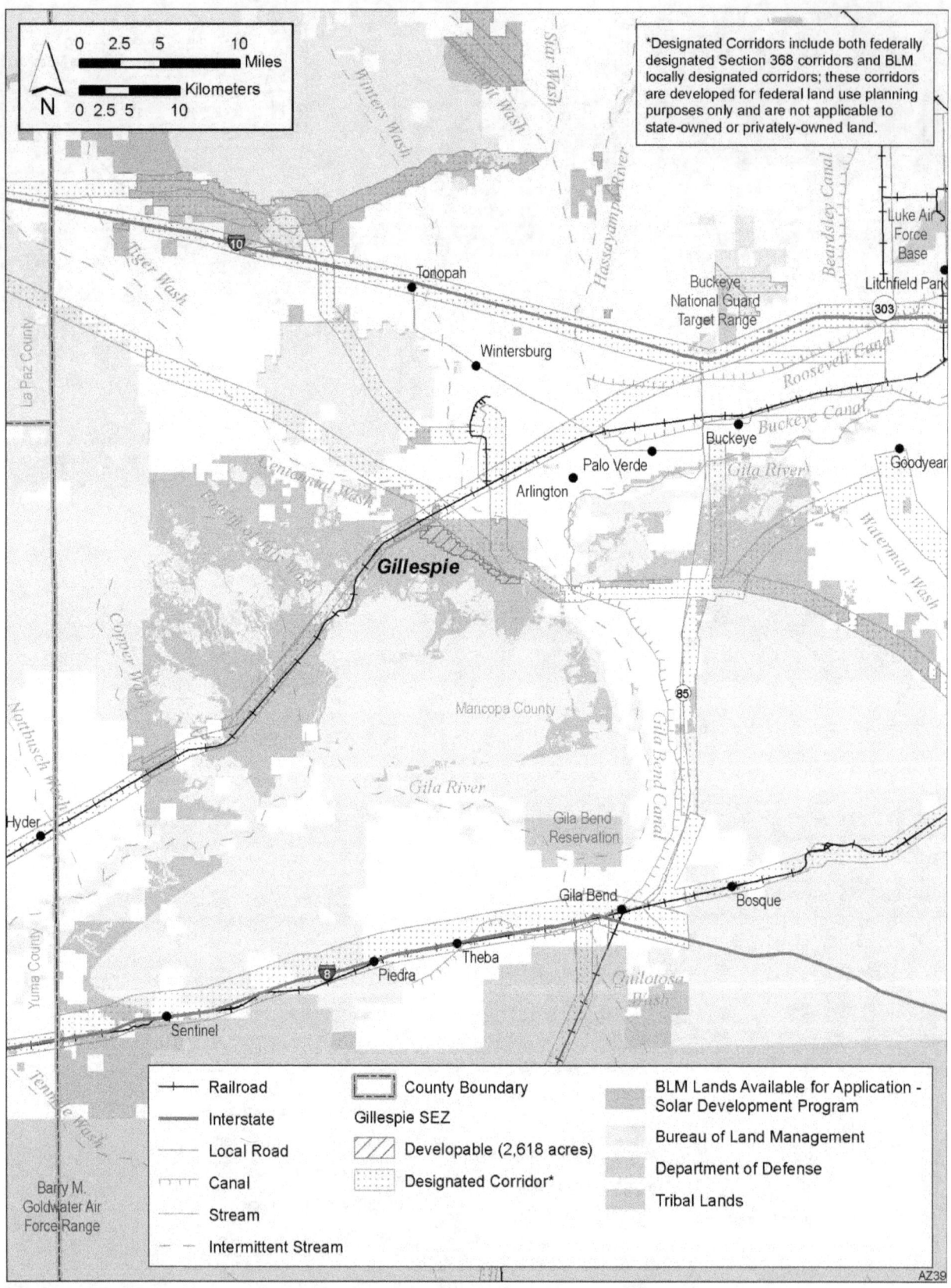

Figure 2-3. Gillespie Solar Energy Zone and surrounding areas as identified in the Solar PEIS ROD

The town of Arlington (population less than 500) is about 7 mi (11 km) northeast of the SEZ, while the larger town of Buckeye is located about 17 mi (27 km) northeast of the SEZ and has a population of more than 50,000. Phoenix, Arizona, is approximately 50 mi (48 km) northeast of the SEZ. Major road access to the SEZ is via Agua Caliente Road, a county road which runs from east to west through the SEZ. The nearest railroad stop is 11 mi (18 km) away. The Palo Verde Nuclear Generating Station, three natural gas power plants, a railroad, transmission lines, and a pipeline right-of-way (ROW) are located in the surrounding area, which has few permanent residences.

2.1.4 Landscape Conditions of the Arizona Solar Energy Zones and the Region

In 2012, the BLM completed the "Sonoran Desert Rapid Ecoregional Assessment" for the Sonoran Desert ecoregion in which the Arizona SEZs are located (BLM 2012c). The Sonoran Desert Rapid Ecoregional Assessment (REA) examines broad-scale ecological values, conditions, and trends within the ecoregion by synthesizing existing spatial datasets in a meaningful time frame. The REAs serve multiple purposes in an ecoregional context, including identifying and answering important management questions; understanding key resource values; understanding the influence of various change agents; understanding projected ecological trends; identifying and mapping key opportunities for resource conservation, restoration, and development; and providing a baseline to evaluate and guide future actions.

One useful product of the REAs is the development of terrestrial landscape intactness models. These geospatial models have been created to represent the level of intactness throughout the ecoregion at the time in which the assessments were initiated (approximately 2010). In the Sonoran Desert REA (BLM 2012c), terrestrial landscape intactness was defined as a quantifiable estimate of naturalness in the ecoregion measured on a gradient of anthropogenic influence (e.g., housing, commercial, and infrastructure development) and was modeled as a function of human development density. The intactness model was developed for a current time period (e.g., 2010 to 2015) as well as a near-term future time period (2025).

Intactness is used as a general indicator of habitat quality based on available spatial data reported at a fairly coarse 4 km X 4 km scale (although current intactness was rescaled to a 1 km x 1 km scale). Figure 2-4 shows current terrestrial intactness of the Sonoran Desert based on the model used in the REA (BLM 2012c). The resulting map provides a composite view of the relative impacts of land uses across the entire ecoregion. Darker green areas indicate the least impacted areas (most intact) and blue areas are the most impacted (least intact). Current terrestrial intactness within the Arizona SEZs is shown in Figure 2-5.

2.1.5 Regional Setting

2.1.5.1 General Description

Agua Caliente SEZ

The Agua Caliente SEZ is located adjacent to the 527,000-acre Yuma East Undeveloped Special Recreation Management Area (SRMA), a BLM-administered area. Other specially designated BLM-administered lands within 20 mi (32 km) of the SEZ include: Sears Point Area of Critical Environmental

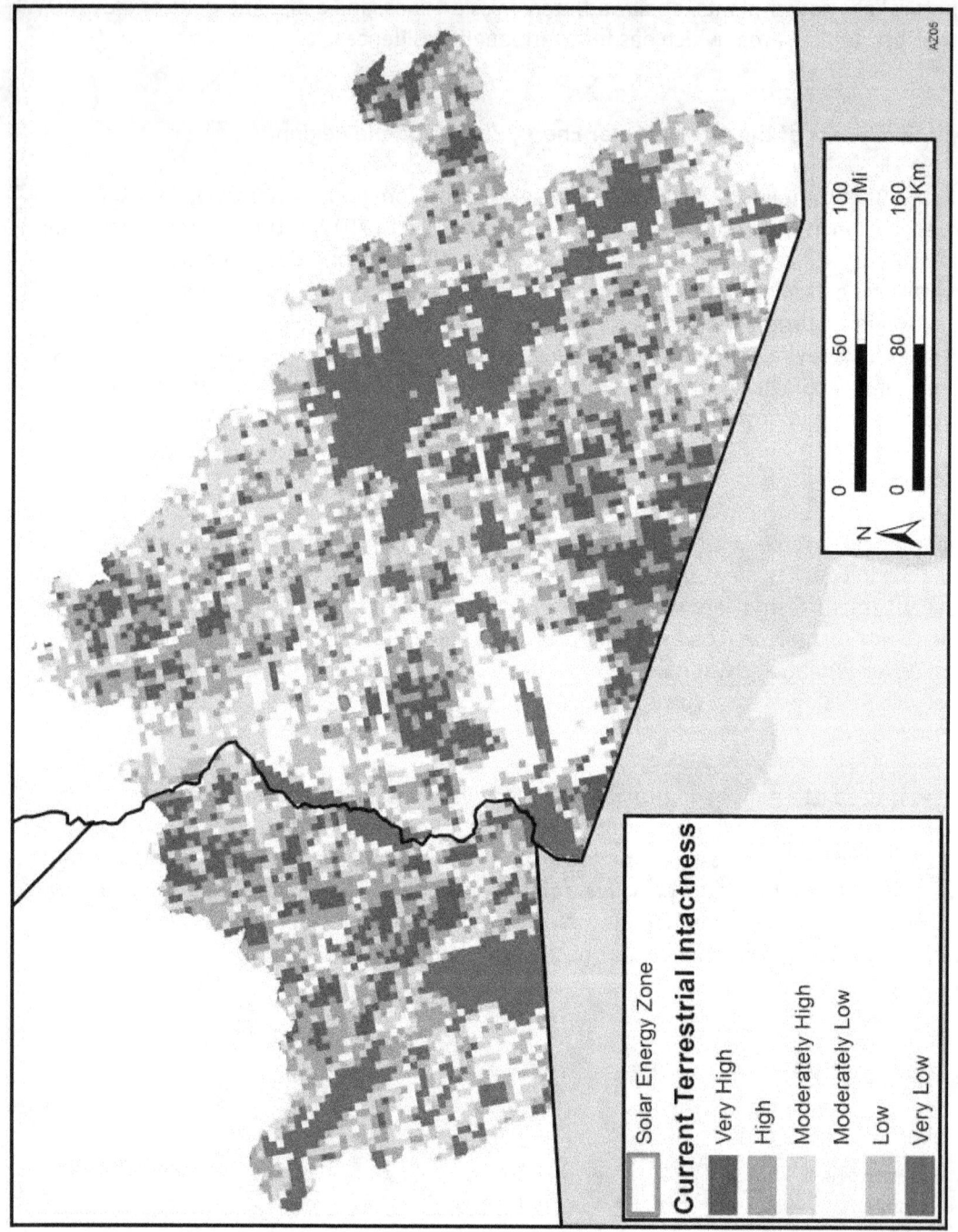

Figure 2-4. Current Sonoran Desert terrestrial landscape intactness in six classes from high (relatively undisturbed in dark green) to very low (highly disturbed from agriculture, resource development, or urbanization in dark blue) depicted with a 4 km X 4 km grid cell (Source: BLM 2012c)

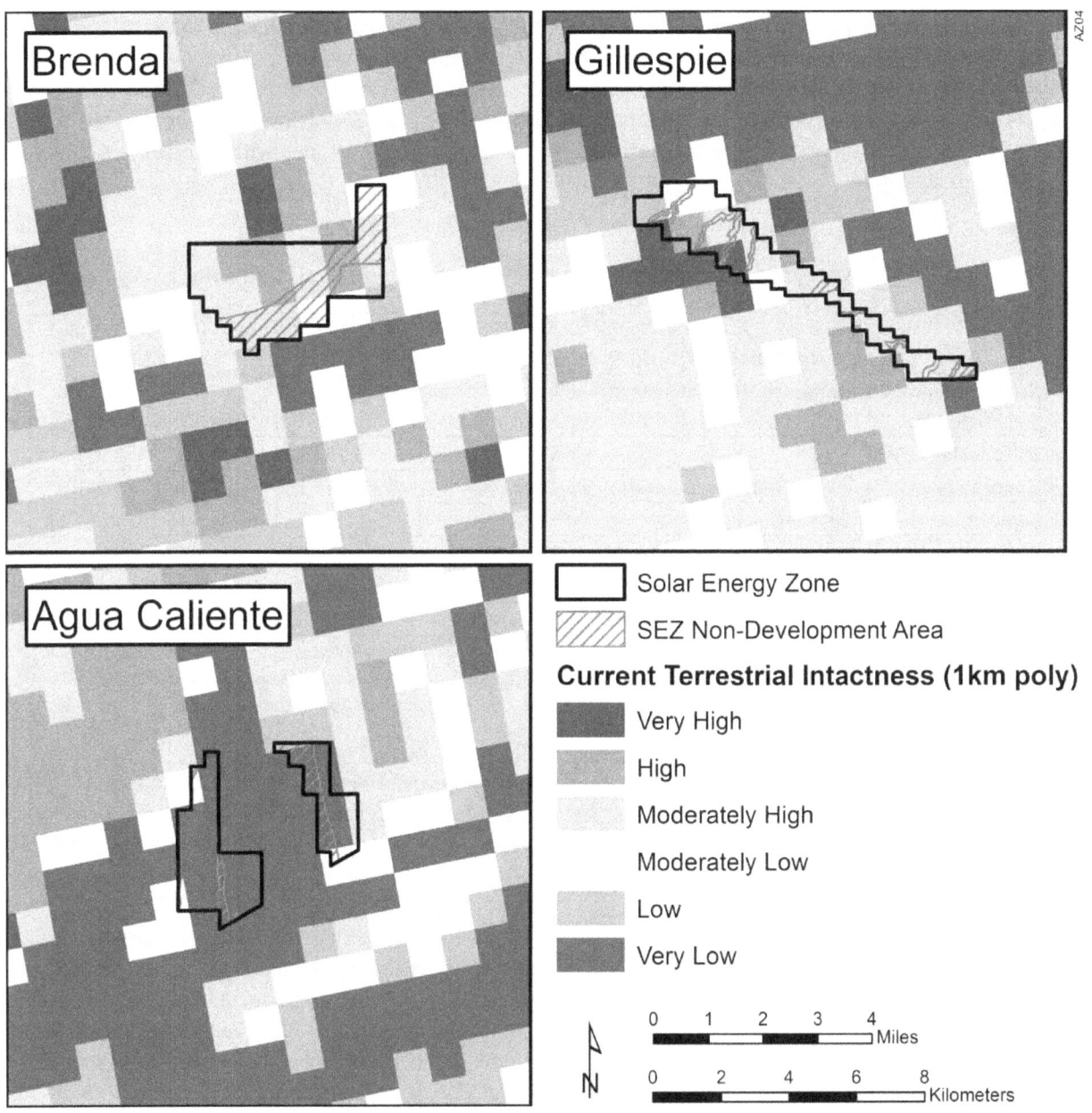

Figure 2-5. Arizona Solar Energy Zones overlain on terrestrial intactness model (Source: BLM 2012c)

Concern (ACEC) (approximately 3.5 mi [5.6 km] from the Agua Caliente SEZ) and the Eagletail Mountains Wilderness Area (15 mi [24 km] from the SEZ). The Juan Bautista de Anza National Historic Trail (NHT), administered by the National Park Service, is located 5 mi (8 km) from the Agua Caliente SEZ.

The Agua Caliente SEZ is undeveloped and rural and is located within the Palomas Plain, which is bounded by the Palomas Mountains to the west and Baragan Mountain to the north. The area surrounding the SEZ contains agricultural lands on the west side, and undeveloped desert to the north, south, and east. In between two parcels of the SEZ is a recently constructed 290-MW photovoltaic solar development on private, previously-agricultural land. The Agua Caliente SEZ is located within the Sonoran Basin and Range ecoregion, dominated by Lower Sonoran desert scrub vegetation.

Based on the distribution of SWReGAP land cover types (USGS National Gap Analysis Program 2004), there are three primary land cover types that occur in the developable portion of the SEZ (Table 2-1), including Sonora-Mojave Creosotebush-White Bursage Desert Scrub (96%), Sonoran Paloverde-Mixed Cacti Desert Scrub (1%), and introduced vegetation(3%). Land cover types in the vicinity of the Agua Caliente SEZ are presented in Figure 2-6. Land cover types within the ecoregion are presented in Figure 2-7.

Table 2-1. Land cover types and amounts in the vicinity of the Agua Caliente Solar Energy Zone

Land Cover Type[1]	Acres Within SEZ Developable Area[2]	Total Acres in SEZ[3]	Acres within 5 mi of SEZ[4]
Sonora-Mojave Creosotebush-White Bursage Desert Scrub	1,902 (95.9%)	2,403 (95.3%)	73,711 (81.9%)
Introduced Vegetation	55 (2.8%)	86 (3.4%)	3,339 (3.7%)
Sonoran Paloverde-Mixed Cacti Desert Scrub	12 (0.6%)	12 (0.5%)	1,977 (2.2%)
Agriculture	11 (0.6%)	15 (0.6%)	10,193 (11.3%)
Barren	2 (0.1%)	2 (0.1%)	47 (0.05%)
North American Warm Desert Riparian Systems	0 (0%)	0 (0%)	578 (0.6%)
Sonora-Mojave Mixed Salt Desert Scrub	0 (0%)	0 (0%)	164 (0.2%)

[1] Data source: SWReGAP land cover types (USGS National Gap Analysis Program 2004).

[2] Values in parentheses represent the percent of acreage relative to the entire developable area (2,021 acres); however, acreages presented in the table may not add up to that exact amount due to rounding and the rasterization of data (fitting boundaries to a grid of 1 km^2 cells).

[3] Values in parentheses represent the percent of acreage relative to the entire SEZ (2,560 acres).

[4] Values in parentheses represent the percent of acreage relative to the entire 5-mile buffer area (90,080 acres).

Brenda SEZ

The Brenda SEZ is situated within 5 mi (8 km) of the Plomosa SRMA. At its nearest point, the Plomosa SRMA is approximately one-eighth of a mile from the western boundary of the SEZ. The SRMA is managed to provide outdoor activities for local residents and visitors. Other specially designated BLM-administered lands within 20 mi (32 km) of the SEZ include: the East Cactus Plain Wilderness Area (20 mi [32 km] north of the SEZ), the Kofa Wilderness Area (14 mi [23 km] south of the SEZ), the New Water Mountains Wilderness Area (6.5 mi [10.5 km] south of the SEZ), the Cactus Plain Wilderness Area (18 mi [29 km] northwest of the SEZ), the Dripping Springs ACEC (9 mi [14 km] from the SEZ). The Kofa National Wildlife Refuge (NWR), managed by the U.S. Fish and Wildlife Service (USFWS) is located approximately 13.5 mi [22 km] from the SEZ.

The Brenda SEZ is undeveloped and rural and is bounded on the north by the Bouse Hills, on the west-southwest by the Plomosa Mountains and the Bear Hills and on the east by the Granite Wash Mountains and Harquahala Mountains. The SEZ is covered by undeveloped scrubland, characteristic of a semi-arid basin desert valley. The Brenda SEZ is located within the Sonoran Basin and Range ecoregion, which supports creosotebush white bursage plant communities with large areas of paloverde cactus shrub and saguaro cactus communities.

Land cover types in the vicinity of the Brenda SEZ are shown in Figure 2-6, and land cover types within the ecoregion are presented in Figure 2-7. In total, there are two land cover types predicted to occur within the Brenda SEZ (with Sonora-Mojave Creosotebush-White Bursage Desert Scrub

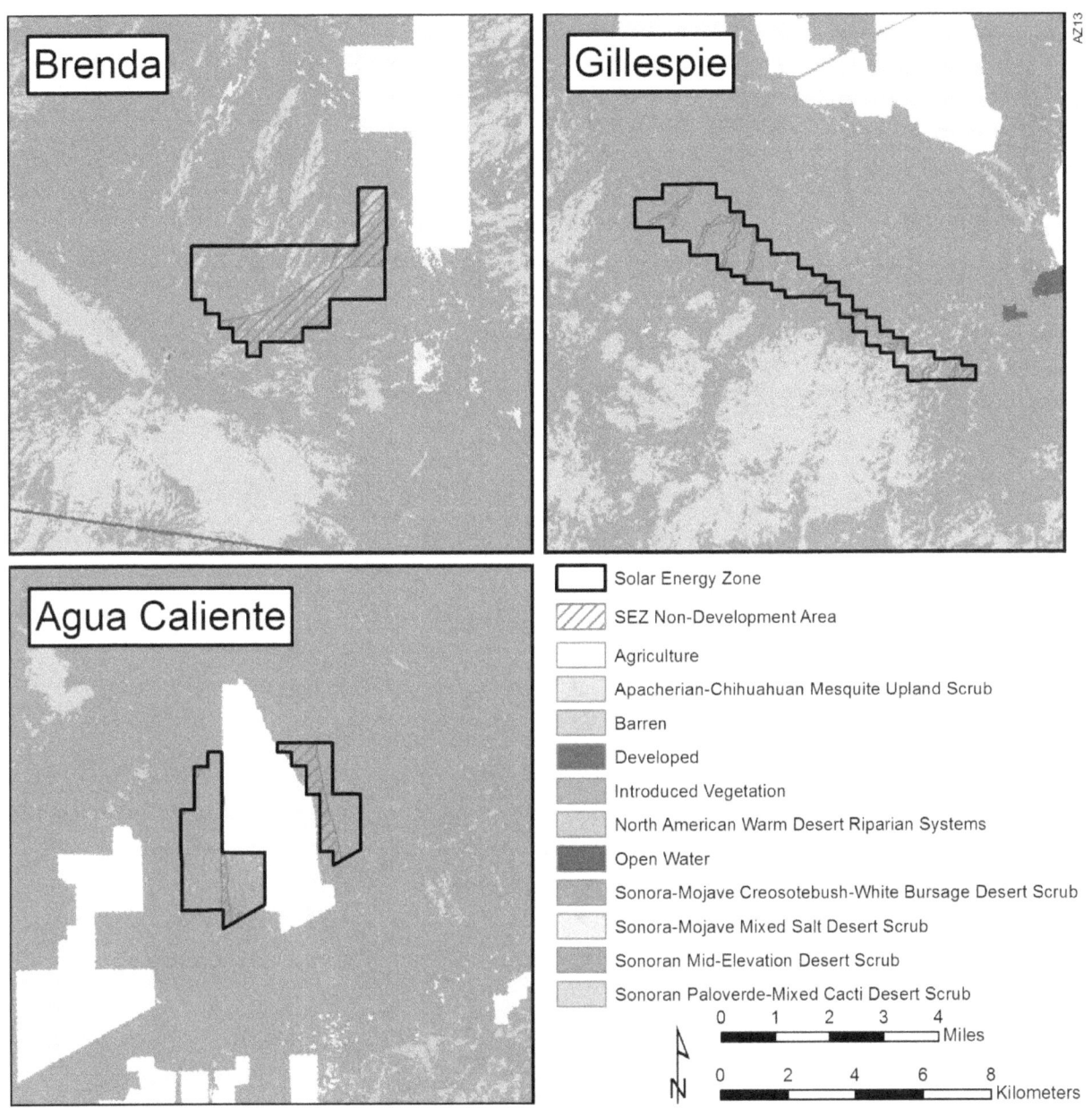

Figure 2-6. Land cover types in the vicinity of the Brenda Solar Energy Zone, Gillespie Solar Energy Zone, and Agua Caliente Solar Energy Zone[5]

---

5   The private land between the Agua Caliente Solar Energy Zone parcels has been developed as a photovoltaic solar facility and is no longer used for agriculture. Land cover databases have not yet been updated to reflect this change. Data Source: SWReGAP Land cover Types (USGS National Gap Analysis Program 2004).

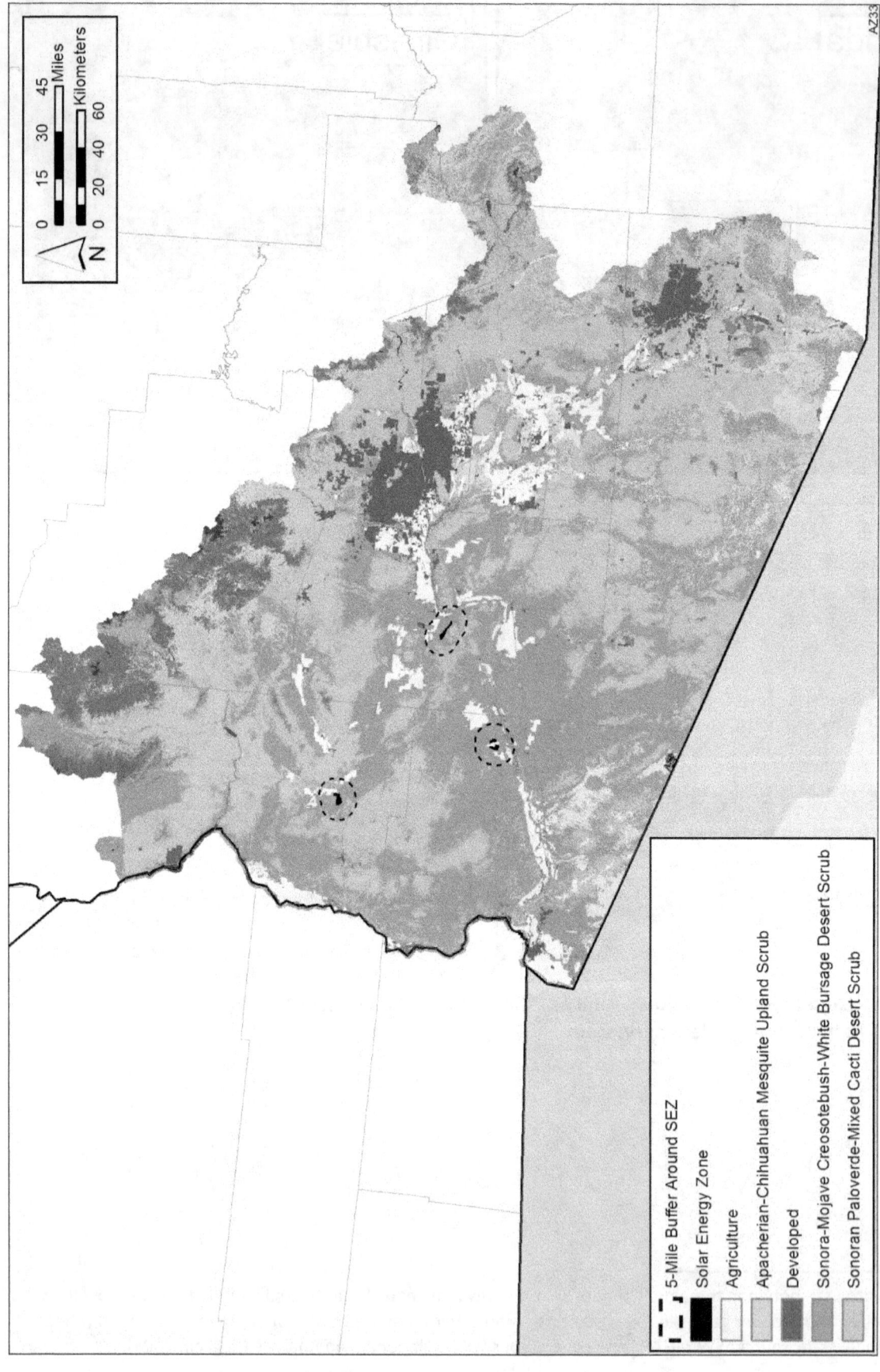

Figure 2-7. Land cover types of the Sonoran Desert ecoregion in Arizona
Source: SWReGAP Landover Types (USGS National Gap Analysis Program 2004)

5-Mile Buffer Around SEZ
Solar Energy Zone
Agriculture
Apacherian-Chihuahuan Mesquite Upland Scrub
Developed
Sonora-Mojave Creosotebush-White Bursage Desert Scrub
Sonoran Paloverde-Mixed Cacti Desert Scrub

comprising 88% of the SEZ), and six land cover types predicted to occur in the vicinity (i.e., within 5 mi, or 8 km) of the SEZ (Table 2-2). Sensitive habitats on the Brenda SEZ include desert dry wash and dry wash woodland.

Table 2-2. Land cover types and amounts in the vicinity of the Brenda Solar Energy Zone

Land Cover Type[1]	Acres Within SEZ Developable Area[2]	Total Acres in SEZ[3]	Acres within 5 mi of SEZ[4]
Sonora-Mojave Creosotebush-White Bursage Desert Scrub	1,720 (91%)	2,920 (87.8%)	58,805 (66.2%)
Sonoran Paloverde-Mixed Cacti Desert Scrub	171 (9%)	407 (12.2%)	22,158 (24.9 %)
Agriculture	0 (0%)	0 (0%)	6,924 (7.8%)
Sonora-Mojave Mixed Salt Desert Scrub	0 (0%)	0 (0%)	541 (0.6%)
Developed	0 (0%)	0 (0%)	309 (0.3%)
Barren	0 (0%)	0 (0%)	72 (0.1%)

[1] Data source: SWReGAP land cover types (USGS National Gap Analysis Program 2004).

[2] Values in parentheses represent the percent of acreage relative to the entire developable area (1,906 acres); however, acreages presented in the table may not add up to that exact amount due to rounding and the rasterization of data (fitting boundaries to a grid of 1 km^2 cells).

[3] Values in parentheses represent the percent of acreage relative to the entire SEZ (3,348 acres).

[4] Values in parentheses represent the percent of acreage relative to the entire 5-mile buffer area (88,001 acres).

Gillespie SEZ

The Gillespie SEZ is situated within 5 mi (8 km) of several BLM-specially designated areas. The Woolsey Peak and Signal Mountain Wilderness Areas are within 2 mi (3 km) and 3.5 mi (6 km), respectively of the boundary of the Gillespie SEZ. Portions of the Saddle Mountain SRMA range from 4–13 mi (6 to 21 km) from the northern boundary of the SEZ. Other specially designated BLM-administered lands within 20 mi (32 km) of the SEZ include: the northwestern portion of Sonoran Desert National Monument (11 mi [18 km] from the SEZ), the North Maricopa Mountains WA (13 mi [21 km] from the SEZ), Eagletail Mountains WA (18 mi [29 km] from the northern boundary of the SEZ), and the Bighorn Mountains and Hummingbird Springs WAs (21 mi [34 km] from the SEZ).

The Gillespie SEZ is undeveloped and rural and is located to the southeast of the Harquahala Basin in a valley between the Gila Bend Mountains to the southwest and Centennial Wash to the northeast. The SEZ is covered by undeveloped scrubland characteristic of a semiarid desert valley. The Gillespie SEZ is located within the Sonoran Basin and Range ecoregion, which supports creosotebush white bursage plant communities with large areas of paloverde cactus shrub and saguaro cactus communities.

Land cover types in the vicinity of the Gillespie SEZ are shown in Figure 2-6, and land cover types within the ecoregion are presented in Figure 2-7. There are two land cover types predicted to occur within the Gillespie SEZ (with Sonora-Mojave Creosotebush-White Bursage Desert Scrub comprising 95% of the SEZ) and six additional land cover types predicted to occur in the vicinity (i.e., within 5 mi, or 8 km) of the SEZ (Table 2-3). Sensitive habitats on the SEZ include desert dry wash and dry wash woodland habitats.

Table 2-3. Land cover types and amounts in the vicinity of the Gillespie Solar Energy Zone

Land Cover Type[1]	Acres Within SEZ Developable Area[2]	Total Acres in SEZ[3]	Acres within 5 mi of SEZ[4]
Sonora-Mojave Creosotebush-White Bursage Desert Scrub	2,067 (94.6%)	2,467 (95.2%)	60,465 (59.4%)
Sonoran Paloverde-Mixed Cacti Desert Scrub	118 (5.4%)	125 (4.8%)	28,324 (27.8%)
Agriculture	0 (0%)	0 (0%)	10,361 (10.2%)
North American Warm Desert Riparian Systems	0 (0%)	0 (0%)	1,564 (1.5%)
Introduced Vegetation	0 (0%)	0 (0%)	447 (0.4%)
Sonora-Mojave Mixed Salt Desert Scrub	0 (0%)	0 (0%)	307 (0.3 %)
Developed	0 (0%)	0 (0%)	238 (0.2%)
Open Water	0 (0%)	0 (0%)	82 (0.1%)

[1] Data source: SWReGAP land cover types (USGS National Gap Analysis Program 2004).

[2] Values in parentheses represent the percent of acreage relative to the entire developable area (2,231 acres); however, acreages presented in the table may not add up to that exact amount due to rounding and the rasterization of data (fitting boundaries to a grid of 1 km^2 cells).

[3] Values in parentheses represent the percent of acreage relative to the entire SEZ (2,618 acres).

[4] Values in parentheses represent the percent of acreage relative to the entire 5-mile buffer area (101,897 acres).

2.1.5.2 Regional Conditions and Trends

The Sonoran Desert REA presents a framework for determining the condition and trend of various resource values and conservation elements in the ecoregion (BLM 2012c). The Sonoran Desert REA defines conservation elements as resources of conservation concern within an ecoregion. These elements could include habitat or populations for plant and animal taxa, such as threatened and endangered species, or ecological systems and plant communities of regional importance. A list of conservation elements could also include other resource values, such as highly erodible soils; scenic viewsheds; or designated sites of natural, historical, or cultural significance. Based on the Sonoran Desert REA, there are three basic types of conservation elements in the Sonoran Desert:

- Coarse filter ecological systems, which represent characteristic vegetation assemblages occurring within the ecoregion.

- Fine filter elements, represented by 11 wildlife species conservation elements as well as a list of designated sites and essential ecosystem functions and services (e.g., aquatic systems, riparian areas, and soil stability).

- Landscape-species conservation elements, in which selected species represent a range of important attributes characterizing the environment in which they occur.

A full list and explanation of ecological systems conservation elements within the Sonoran Desert can be found in Appendix B of the Sonoran Desert REA (BLM 2012c). Examples of fine-filter plant species include saguaro and creosote bush. For landscape-species conservation elements, Sonoran desert tortoise, lowland leopard frog, and Le Conte's thrasher are examples.

Problematic trends are understood by forecasting the response of conservation elements to one of four change agents in the ecoregion. The four change agents include fire, invasive species, climate change, and human development. Of these change agents, the conservation element vulnerability to human development and climate change are used in this assessment to evaluate resource conditions and trends.[6]

Understanding the conservation element trends relevant to the Arizona SEZs was accomplished through (1) a geospatial analysis of available ecoregional data, (2) expert opinion by a BLM interdisciplinary team, and (3) comments from knowledgeable stakeholders. The geospatial data used in this assessment are available publicly from open sources. These data include the BLM's terrestrial landscape intactness model for the Sonoran Desert (see Section 2.1.4), SWReGAP modeled land cover types (USGS National Gap Analysis Program 2004), and species-specific habitat suitability models. The Sonoran Desert terrestrial landscape intactness model can be used to represent regional landscape intactness. Evaluating condition and trends of coarse and fine filter conservation elements (land cover and habitat models) in an ecoregional context will provide a better understanding of the impacts of solar energy development within the Arizona SEZs relative to the rest of the ecoregion.

The geospatial process for quantitatively evaluating condition and trends for conservation elements begins with a characterization of the distribution of the conservation element within identified analysis areas: (1) the entire Sonoran Desert ecoregion, (2) the vicinity of the Arizona SEZs, and (3) within each of the Arizona SEZ developable areas. These areas are then clipped to current and anticipated future ecological intactness models and climate forecast trends. Due to the coarse scale of some of the REA datasets (e.g., 4 km^2) and the relatively small size of the SEZs, the BLM determined that condition and trend assessments for individual species would not be appropriate for this SRMS. Instead, general measures of condition and trend for the ecological systems of each SEZ were evaluated using the following REA datasets: current and future intactness models, ecological systems, solar energy development data, and climate change models. General landscape conditions and trends for each of the three SEZs relative to the Sonoran Ecoregion are shown in Table 2-4.

Based on the information presented in Table 2-4, it was concluded that ecological systems for all three SEZs are expected to experience a declining trend in the Sonoran Desert, as solar development on the SEZs is expected to contribute to a loss of ecological intactness on and in the vicinity of the SEZs. The terrestrial landscape intactness within the Sonoran Desert is expected to decline in the future, as well as in the three SEZs. Because the Sonoran-Mojave Creosotebush-White Bursage Desert Scrub and Sonoran Paloverde-Mixed Cacti Desert Scrub vegetation communities comprise the largest portions of the Arizona SEZs (Agua Caliente 95%, 1%; Brenda 91%, 9%; and Gillespie 95%, 5%; respectively), the cumulative expected future loss or degradation of these ecological systems due to human development and climate change is considered to be an important trend in the region for these ecological systems and other conservation elements.

---

[6] Conditions and trends of Conservation Elements evaluated in this SRMS considered the human development (including agriculture and grazing) and climate change REA change agents. These two change agents are fundamental drivers of landscape change as they influence, at least in part, the other two change agents (i.e., invasive species and wildfire).

Table 2-4. Ecological intactness of the Arizona Solar Energy Zones and condition assessment for ecological systems

A) Average intactness within SEZs relative to the ecoregion

SEZ	Current Ecological Intactness (4 km²)			Future Ecological Intactness (4 km²)[1]		
	Average Intactness within SEZ Dev. Area	Average Intactness within Buffer	Average Intactness within Ecoregion	Average Intactness within SEZ Dev. Area	Average Intactness within Buffer	Average Intactness within Ecoregion
Agua Caliente	Low (-0.68)	Low (-0.62)	Moderately High (0.10)	Very Low (-0.92)	Low (-0.73)	Moderately High (0.07)
Brenda	Moderately High (0.38)	Moderately Low (-0.01)		Very Low (-1.00)	Very Low (-0.82)	
Gillespie	Moderately Low (-0.05)	Moderately Low (-0.21)		Very Low (-0.83)	Low (-0.62)	

Source: Sonoron Desert Rapid Ecoregional Assessment (BLM 2012c)

B) Average intactness of the ecological systems on the SEZs and throughout the Sonoran Desert ecoregion in Arizona.

Ecological System	SEZ Distribution				Ecoregional Condition Assessment		
	Ecoregional Distribution within AZ (%)	Percent Within Agua Caliente SEZ	Percent Within Brenda SEZ	Percent Within Gillespie SEZ	Average Current Intactness	Average Future Intactness	Potential for Future Climate Change[2]
Sonora-Mojave Creosotebush-White Bursage Desert Scrub	31%	96%	91%	95%	Moderately High	Moderately Low	Very High
Sonoran Paloverde-Mixed Cacti Desert Scrub	35%	1%	9%	5%	Very High	Moderately High	Moderately Low

Data sources: USGS National Gap Analysis Program 2004; BLM 2012c.
[1] Future ecological intactness assumes full development of the SEZs.
[2] Climate change models developed for the REA were based on predicted future seasonal changes in precipitation and temperature.

2.2 General Description of Solar Development in the Arizona Solar Energy Zones

2.2.1 Description of Existing Rights-of-Way, Development Status, and Recommended Non-Development Areas

2.2.1.1 Agua Caliente Solar Energy Zone

The Agua Caliente SEZ is undeveloped, and the area around the SEZ is sparsely populated. Numerous transportation routes traverse the SEZ, most heading north-south and crossing or originating/terminating on private and state land.

There are no current applications for solar development within the Agua Caliente SEZ, and there is one pending application located within a 25-mile (40-km) radius of the SEZ. A 290-MW Photovoltaic (PV) facility located on private lands adjacent to the SEZ began operations in 2014. The facility is owned by NRG Energy and the electricity is being sold to Pacific Gas and Electric Company, which serves citizens primarily in Southern California.

Since the signing of the RDEP ROD, BLM has collected, compiled, and/or reviewed new data and analyses and recommends avoidance of Federal Emergency Management Agency (FEMA) floodplain areas, cultural resources identified during an archaeological survey of the entire SEZ, and lands with wilderness characteristics. Of the 2,560 acres (10.3 km^2) of developable area, the SRMS recommends development on up to 2,021 acres (8.18 km^2) (Figure 2-8). Non-development areas would be finalized during pre-auction NEPA analysis (see Figure 1-1) with the intention that they would not be made available during auction.

2.2.1.2 Brenda Solar Energy Zone

The Brenda SEZ is rural and undeveloped. The community of Brenda is located about 3 mi (5 km) southwest of the SEZ. There is land disturbance to the south and west of the SEZ associated with road construction, power line construction, mining, and development of the town site. There are scattered home sites and RV parks along U.S. 60. A 500-kV transmission line is located approximately 12 mi (19 km) south of the SEZ, and there is a designated transmission corridor adjacent to the southern SEZ boundary.

There are no pending solar project applications within the SEZ. There is one authorized, but unconstructed, solar project (the Quartzsite Solar Energy Project, a 100-MW power tower facility) located about 15 mi [24 km] northwest of the SEZ and one pending solar facility application (the Little Horn project) located about 20 mi [32 km] southeast of the SEZ within a 25-mile (40-km) radius of the SEZ.

Since the signing of the Solar PEIS ROD, BLM has collected, compiled, and/or reviewed new data and analyses and recommends avoidance of FEMA floodplain areas (data not available at time of Solar PEIS), cultural resources identified during an archaeological survey of the entire SEZ, and other sensitive resources. Of the 3,348 acres (13.5 km^2) of developable area, the SRMS recommends development on up to 1,906 acres (7.71 km^2) (Figure 2-9). Non-development areas would be finalized during pre-auction NEPA analysis (see Figure 1-1) with the intention that they would not be made available during auction.

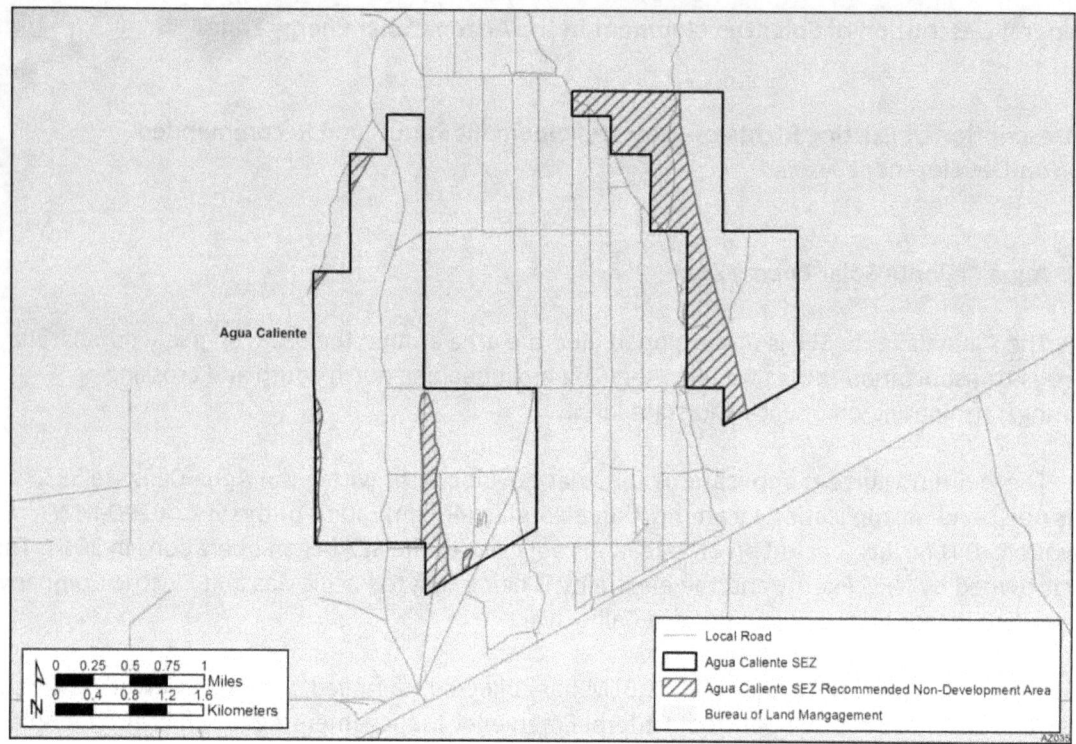

Figure 2-8. Agua Caliente Solar Energy Zone recommended developable area

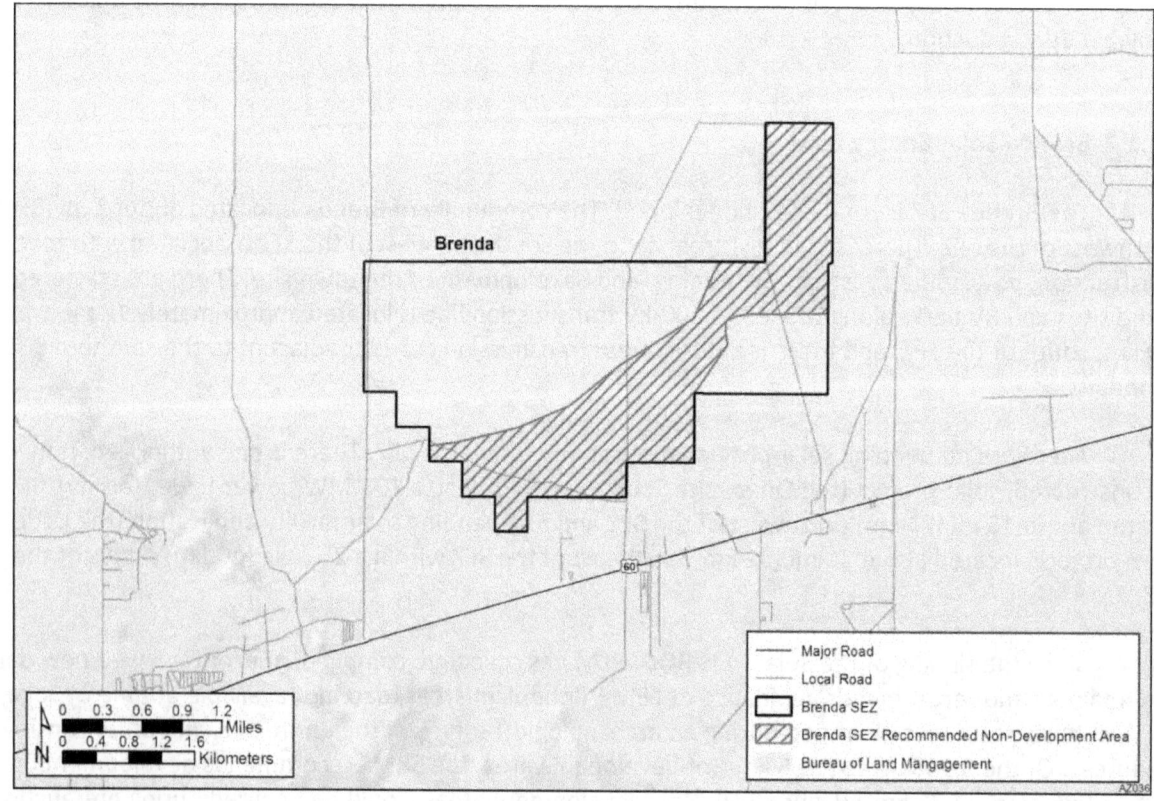

Figure 2-9. Brenda Solar Energy Zone recommended developable area

2.2.1.3 Gillespie Solar Energy Zone

The Gillespie SEZ is rural and undeveloped. The area is used primarily for grazing and some recreational activities. The Palo Verde Nuclear Generating Station is located about 6 mi (10 km) north of the SEZ, and two large capacity transmission lines pass within 0.5 mi (0.8 km) of the SEZ. These transmission lines are located within designated local ROW corridors, and portions of these local corridors also have been identified as 368(b) corridors. A branch of the Union Pacific Railroad passes along the northwestern edge of the SEZ.

There are no pending solar project applications within the SEZ, and there is one authorized, but unconstructed solar project (the Sonoran Solar Energy Project, a 300-MW photovoltaic facility) located about 12 mi [19 km] east of the SEZ. There are no pending solar applications within a 25-mile (40-km) radius of the SEZ.

Since the signing of the Solar PEIS ROD, BLM has collected, compiled, and/or reviewed new data and analyses and recommends avoidance of FEMA floodplain areas (data not available at time of Solar PEIS), cultural resources identified during an archaeological survey of the entire SEZ, and other sensitive resources. Of the 2,618 acres (11 km^2) of developable area, the SRMS recommends development on up to 2,231 acres (9.03 km^2) (Figure 2-10). Non-development areas would be finalized during pre-auction NEPA analysis (see Figure 1-1) with the intention that they would not be made available during auction.

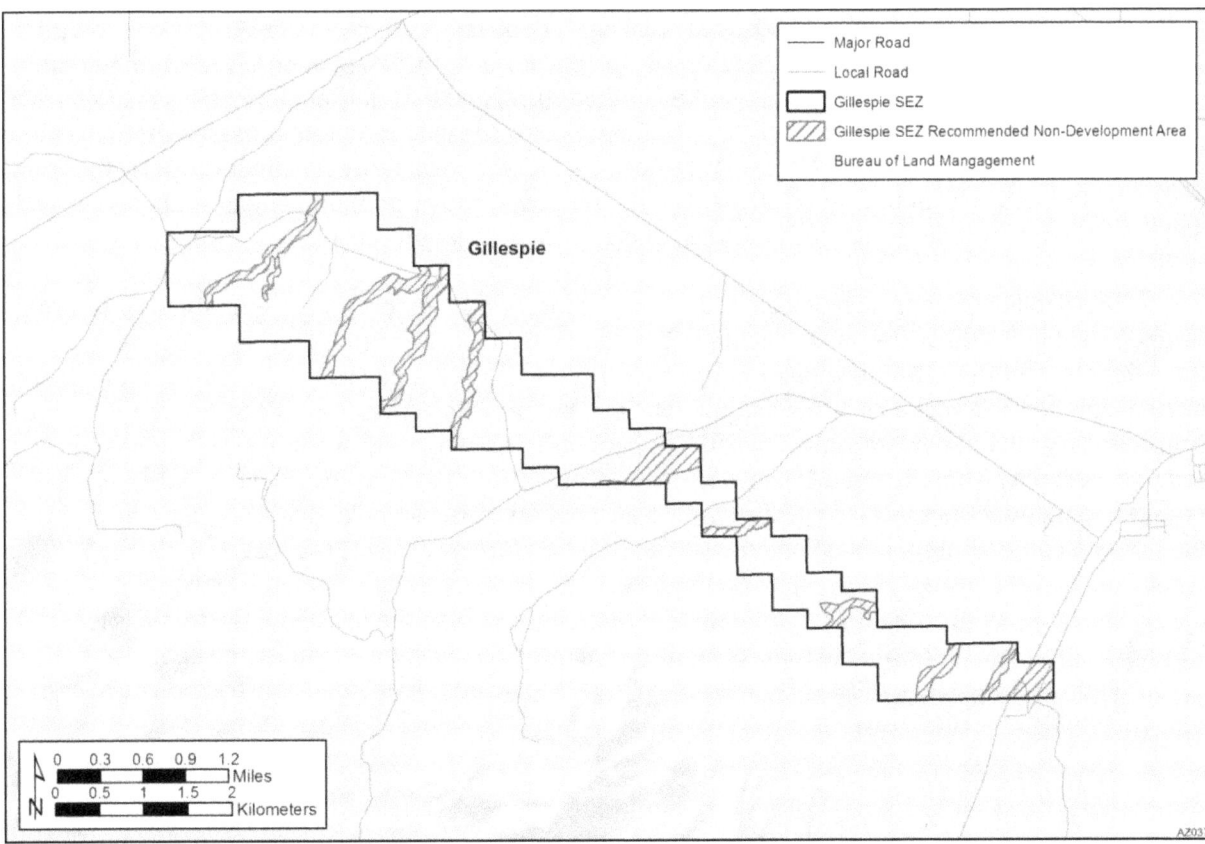

Figure 2-10. Gillespie Solar Energy Zone recommended developable area

2.2.2 Description of Potential Development

Utility-scale solar facilities of all technology types have a key element in common—they all have a large solar field with reflectors or photovoltaic surfaces designed to capture the sun's energy. The solar fields generally require a relatively flat land surface; only locations with less than 5% slope were included as SEZs in the Final Solar PEIS. As typically constructed to date and as assumed in the Solar PEIS for determining impacts, vegetation is generally cleared from solar fields prior to construction, and the fields are fenced to prevent damage to or from wildlife and trespassers. However, alternative site preparation methods may limit vegetation clearing and corresponding impacts to soil, vegetation, and dust generation.

In the Final Solar PEIS, maximum solar development of the SEZs was assumed to be 80% of the developable SEZ area over a period of 20 years. Data from various existing solar facilities were used to estimate that solar trough facilities will require about 5 acres/megawatt (0.02 km2/megawatt), and other types of solar facilities (e.g., power tower and photovoltaic technologies) will require about 9 acres/megawatt (0.04 km2/megawatt).[7]

Agua Caliente SEZ

Although the size of the developable area for the Agua Caliente SEZ was reduced in the RDEP ROD as an avoidance measure (see Section 2.2.1), the BLM recommends avoiding FEMA floodplain areas, cultural resources identified during an archaeological survey of the entire SEZ, and lands with wilderness characteristics. Of the 2,560 acres (10.3 km2) of developable area, the SRMS recommends development on up to 2,021 acres (8.18 km^2). For the purposes of this assessment, it is assumed that more non-development areas may be identified in the future, and that only about 1,617 acres (6.5 km^2) (80% of the recommended acreage) will be developed. Using the land requirement assumptions described above, full development of the Agua Caliente SEZ would allow development of solar facilities with an estimated total of between 180 megawatts (for power tower or photovoltaic technologies) and 323 megawatts (for solar trough technologies) of electrical generation capacity.

Availability of transmission from SEZs to load centers is an important consideration for future development in SEZs. For the Agua Caliente SEZ, a 500-kV east-west transmission line is located 0.5 miles south of the SEZ. A new Hassayampa to North Gila 500-kV transmission line was put into service in 2015. It is possible that an existing line could be used to provide access from the SEZ to the transmission grid, but new transmission and/or upgrades of existing transmission lines may be required to bring electricity from the Agua Caliente SEZ to load centers. Project-specific analyses would also be required to identify the specific impacts of new transmission construction and line upgrades and appropriate compensatory mitigation for that transmission for any projects proposed within the SEZ.

A Yuma County road (Palomas Road) that provides access to the SEZ and Interstate 8 is located 12 mi (19 km) south of the SEZ; therefore, existing road access should be adequate to support construction and operation of solar facilities. It is likely that no additional road construction outside of the SEZ would be needed.

---

[7] Development requirements for low slope and resource impact avoidance (e.g., visual resources, ephemeral streams) will likely limit the technology options for the Arizona SEZs. The most likely technology to be implemented at all three SEZs is photovoltaic.

Brenda SEZ

The Brenda SEZ is 3,348 acres (13.5 km^2), as established in the Solar PEIS ROD. The BLM recommends avoiding newly identified resource conflicts, limiting development on up to 1,906 acres (7.71 km^2). For the purposes of this assessment, it is assumed that more non-development areas may be identified in the future, and that only about 1,525 acres (6.2 km^2) (80% of the recommended acreage) will be developed. Using the land requirement assumptions described above, full development of the Brenda SEZ would allow development of solar facilities with an estimated total of between 169 megawatts (for power tower or photovoltaic technologies) and 305 megawatts (for solar trough technologies) of electrical generation capacity.

For the Brenda SEZ, a 500-kV transmission line passes 12 mi (19 km) south of the SEZ. It is possible that a new transmission line could be constructed from the SEZ to the existing line, but the available capacity on the existing 500-kV could be inadequate for the new capacity. Therefore, new transmission and/or upgrades of existing transmission lines would likely be required to bring electricity from the Brenda SEZ to load centers. An assessment of the most likely load center destinations for power generated at the Brenda SEZ and a general assessment of the impacts of constructing and operating new transmission facilities on those load centers was provided in Section 8.1.23 of the Final Solar PEIS. Project-specific analyses would also be required to identify the specific impacts of new transmission construction and line upgrades for any projects proposed within the SEZ.

Since U.S. 60 runs southwest to northeast along the southeast border of the Brenda SEZ, existing road access should be adequate to support construction and operation of solar facilities. It is likely that no additional road construction outside of the SEZ would be needed.

Gillespie SEZ

The Gillespie SEZ is 2,618 acres (10.6 km^2), as established in the Solar PEIS ROD. The BLM recommends avoiding newly identified resource conflicts, limiting development on up to 2,231 acres (9.0 km^2). For the purposes of this assessment, it is assumed that more non-development areas may be identified in the future, and that only about 1,785 acres (7.2 km^2) (80% of the recommended acreage) will be developed. Using the land requirement assumptions described above, full development of the Gillespie SEZ would allow development of solar facilities with an estimated total of between 198 megawatts (for power tower or photovoltaic technologies) and 357 megawatts (for solar trough technologies) of electrical generation capacity.

For the Gillespie SEZ, a 500-kV transmission line runs less than 1 mi (1.6 km) west of the SEZ. It is possible that an existing line could be used to provide access from the SEZ to the transmission grid, but since existing lines may already be at full capacity, it is possible that new transmission and/or upgrades of existing transmission lines may be required to bring electricity from the Gillespie SEZ to load centers. An assessment of the most likely load center destinations for power generated at the Gillespie SEZ and a general assessment of the impacts of constructing and operating new transmission facilities on those load centers was provided in Section 8.3.23 of the Final Solar PEIS. Project-specific analyses would also be required to identify the specific impacts of new transmission construction and line upgrades for any projects proposed within the SEZ.

Agua Caliente Road runs through the Gillespie SEZ so existing road access should be adequate to support construction and operation of solar facilities. It is likely that no additional road construction outside of the SEZ would be needed.

2.3 Summary of Solar Development Impacts on the Arizona Solar Energy Zones

Comprehensive assessment of the potential impacts of solar development at the Brenda SEZ and Gillespie SEZ was provided in the Final Solar PEIS (BLM and DOE 2012). Assessment of the potential impacts at the Agua Caliente SEZ was provided in the Final RDEP EIS (BLM 2012d). Identified potential adverse impacts included effects on nearby wilderness areas, recreational use of the SEZ lands, military use of the airspace over the SEZ lands, soils, water resources, vegetation, wildlife, special status species (both vegetation and wildlife), air quality, visual resources, paleontological and cultural resources, Native American concerns, and transportation. Some potential positive impacts of development were identified for local socioeconomics due to an increase in local employment and in the reduction of greenhouse gas emissions if solar energy produced at the SEZs would displace use of fossil fuels.

2.4 Mitigation Strategy (Hierarchy) for the Arizona Solar Energy Zones

2.4.1 Avoidance

Agua Caliente SEZ

As discussed in Section 2.1.1, the boundaries of the Agua Caliente SEZ were revised in the Final RDEP EIS to exclude major washes, maintain an area for potential tortoise migration between the Palomas Mountains and Baragan Mountain, and avoid most known archaeological sites and lands with wilderness characteristics (BLM 2013a). On the basis of new information on floodplains, cultural resources, and lands with wilderness characteristics (see Section 2.2.1.1), the BLM recommends avoiding these resources. Avoidance of these areas will also reduce potential impacts identified in the Final RDEP EIS (e.g., fewer acres of habitat reduction will occur for vegetation and wildlife species, including special status species).

Brenda SEZ

As discussed in Section 2.1.1, the boundaries of the Brenda SEZ were revised in the Solar PEIS ROD to eliminate Bouse Wash and an area on the west side of the SEZ (BLM 2012a). On the basis of new information on floodplains, cultural resources, and other sensitive resources (see Section 2.2.1.2), BLM recommends avoiding these resources. Avoidance of these areas will reduce potential impacts identified in the PEIS (e.g., fewer acres of habitat reduction will occur for vegetation and wildlife species, including special status species).

Gillespie SEZ

As discussed in Section 2.2.1.3, the BLM recommends avoidance of some areas on the basis of new information on floodplains, cultural resources, and other sensitive resources. Avoidance of these areas will reduce potential impacts identified in the PEIS (e.g., fewer acres of habitat reduction will occur for vegetation and wildlife species, including special status species).

2.4.2 Minimization

2.4.2.1 Summary of Programmatic Design Features to be Applied

The Solar PEIS ROD and the RDEP ROD identified a comprehensive suite of required programmatic design features that would avoid and/or minimize adverse impacts to resources, either onsite or through consultation and/or coordination with potentially affected entities. The programmatic design features are extensive and are listed in their entirety in Appendix A of the Solar PEIS ROD (BLM 2012a). These programmatic design features, which will be applied to solar development on BLM-administered lands in Arizona, include required actions to avoid or minimize impacts to all of the potentially impacted resources listed in Section 2.3.

2.4.2.2 Other Required Impact Minimization Measures and/or Stipulations

The Solar PEIS ROD also identifies SEZ-specific design features. The SEZ-specific design features identified for the Brenda and Gillespie SEZs are listed below. Some SEZ-specific design features were also identified for the Agua Caliente SEZ in the RDEP ROD.

Agua Caliente SEZ

Water resources: The SEZ is located in Water Protection Zone 2. Industrial water use is limited to solar photovoltaic, solar thermal with dry-cooling, or similar low-water use technologies.

Wildlife (Mammals): Report sightings of or signs of Sonoran pronghorn in the vicinity of the Agua Caliente SEZ to the USFWS and Arizona Game and Fish Department. Documentation of sightings of animals, tracks, droppings, and hair, through digital or other photography, to the extent practical, is recommended.

Wildlife (Mammals): Lay out of fencing around renewable energy facilities avoid creating "dead end" or "trap" areas between fenced areas to allow easy egress for Sonoran pronghorn from the area if startled by humans or predators. The USFWS also recommended designing fencing to avoid ensnarling pronghorn and other large mammals.

Wildlife (Mammals): Include briefing materials on Sonoran pronghorn in Worker Education and Awareness Programs for construction workers at renewable energy facility sites within the Agua Caliente SEZ, including identification and the importance of avoiding disturbing any animals encountered. The USFWS also recommended that the BLM work with them and the Arizona Game and Fish Department in development of Worker Education and Awareness Programs material for Sonoran pronghorn.

Wildlife (Mammals): Keep work areas clean, including eliminating edible garbage and prohibiting the feeding of animals.

Brenda SEZ

Water resources: Groundwater analyses suggest that full build-out of wet-cooled technologies is not feasible; for mixed-technology development scenarios, any proposed wet-cooled projects would be required to employ water conservation practices. Per the RDEP ROD, the SEZ is located in Water Protection Zone 3 and new water uses and withdrawals are restricted to panel washing and sanitary uses only.

Acoustics: Because of the proximity of the SEZ to nearby residences and the Plomosa SRMA and the relatively high noise levels around the SEZ due to U.S. 60, refined modeling would be warranted along with background noise measurements during project-specific assessments.

Gillespie SEZ

Lands and Realty: Priority consideration should be given to using the existing Agua Caliente Road to provide construction and operations access to the SEZ. Any potential impacts on the existing county road should be discussed with the county.

Recreation: Because of the potential for solar development to sever current access routes departing the county road within the SEZ, legal access to the areas to the south should be maintained consistent with existing land use plans.

Water resources: Groundwater analyses suggest that full build-out of wet-cooled technologies is not feasible; for mixed-technology development scenarios, any proposed wet-cooled projects would be required to employ water conservation practices. Per the RDEP ROD, the SEZ is located in Water Protection Zone 3 and new water uses and withdrawals are restricted to panel washing and sanitary uses only (BLM 2013a).

Wildlife (Mammals): The fencing around the solar energy development should not block the free movement of mammals, particularly big game species.

Visual resources: Due to potential visual impacts on two Wilderness Areas, visual impact mitigation should be considered for any solar development within the SEZ.

Cultural resources: Recordation of historic structures through Historic American Building Survey/Historic American Engineering Record protocols through the National Park Service would be appropriate and could be required if any historic structures or features would be affected; for example, if the Gillespie Dam Highway Bridge were used as part of an off-site access route for a solar energy project.

2.4.3 Regional Compensatory Mitigation

Identifying the impacts of utility-scale solar development that may warrant regional compensatory mitigation involves three steps: (1) identifying all of the potential impacts; (2) identifying which of the potential impacts are likely to be residual impacts (i.e., that cannot be avoided or minimized); and (3) identifying which of the residual impacts may warrant regional compensatory mitigation by taking into consideration the condition and trend of the impacted resources in the region in the context of existing policy and law regarding those resources and how that condition and trend could be affected by the residual impacts.

As part of the SRMS process, a team of specialists from the BLM Renewable Energy Coordination Office within the Arizona State Office (called the interdisciplinary team, or IDT) reevaluated the potential impacts of solar development that were described in the Final Solar PEIS (see Section 2.3) in the light of available data specific to the SEZ areas. This team, along with other subject matter experts from the local BLM field offices in which the SEZs are located and from Argonne National Laboratory, followed the methodology presented in Sections 2.4.3.1 and 2.4.3.2 for first identifying residual impacts from solar development in the SEZ, and then for identifying the residual impacts that may warrant regional compensatory mitigation. The identification of residual impacts and residual impacts that may warrant regional compensatory mitigation was presented to the public and their input was incorporated into this draft SRMS.

2.4.3.1 Identification of Residual Impacts

The following methodology was used to identify residual impacts:

- The IDT verified and/or augmented the affected environment and impacts presented in the Final Solar PEIS (for completeness, staff reviewed analyses in both the Draft and Final Solar PEIS) and the Final RDEP EIS.
    - The IDT reviewed the affected environment and the direct, indirect, and cumulative impacts for each resource value presented in the Final Solar PEIS and Final RDEP EIS.
    - The IDT evaluated whether the description of the affected environment and impacts was comprehensive and accurate and whether more detailed information was available that could influence the description of impacts as provided in the Final Solar PEIS and Final RDEP EIS.

- The IDT verified and/or augmented the programmatic and SEZ-specific design features presented in Appendix A of the Solar PEIS ROD and RDEP ROD.
    - The IDT reviewed the programmatic and SEZ-specific design features presented in the Solar PEIS ROD, determined which design features are applicable to the Arizona SEZs, and determined if there are additional measures that could be implemented to avoid and/or minimize impacts. Where applicable, these additional mitigation measures are documented as requiring evaluation in project-specific NEPA in Appendix A.

- The IDT identified the impacts that could be mitigated through avoidance and/or minimization, assuming the required design features described previously would be implemented.
    - For each resource, the design features were evaluated by the IDT as to the degree that they could avoid and minimize the impacts.

- The IDT identified the residual impacts (i.e., those impacts that would remain after implementation of required design features).

The summary tables presented in Appendix A document the basis for the identification of residual (unavoidable) impacts for the Arizona SEZs.

2.4.3.2 Residual Impacts that May Warrant Regional Compensatory Mitigation

2.4.3.2.1 Conceptual Models

A conceptual model (or models) depicting interrelationships between key ecosystem components, processes, and stressors at the Arizona SEZs is fundamental to understanding impacts and for evaluating the effectiveness of compensatory mitigation investments employed through an SRMS. The AZ SRMS team constructed conceptual models to explain the role that resources, individually and in concert with one another, play in the function of the relevant ecological, social, and cultural systems present in the region. This regional model provided the context to identify critical resources at the local scale. Information sources used for the development of the conceptual models included:

- Sonoran Desert REA (BLM 2012c)

- Yuma RMP, Lake Havasu RMP, Lower Sonoran RMP (BLM 2010, BLM 2007, BLM 2012b)

- Resource specialist expert opinion

Additional resources (e.g., other baseline resource surveys, inventories, occurrence records, research studies, assessments, and plans providing insight into regional conditions and trends; ethnographic studies; county or regional land use plans; and federal, state, or local social and economic studies) could be used to refine the models in the future.

Conceptual models for the Sonoran Desert ecosystem, for solar energy development, and for solar energy development at the Arizona SEZs were developed with a goal of describing in detail the processes essential to sustain the ecosystem and the stressors that influence those processes. These conceptual models are presented in Appendix B.

2.4.3.2.2 Residual Impacts Warranting Compensatory Mitigation

On the basis of the list of residual impacts identified (Sec. 2.4.3.1.) and the best available information, conceptual models, assessments, and expert opinion, the Arizona IDT identified those residual impacts that may warrant compensatory mitigation in the context of existing policy and laws and current resource management plans' goals and objectives regarding those resources. The IDT analyzed how the residual impacts of solar development, at full build-out in the SEZs, could affect the condition and trend of the resource values at both local and regional scales. The following criteria were also considered in determining if compensatory mitigation may be warranted:

a. The relative importance placed on the resource in the land use plan.

b. The rarity, legal status, or state or national policy status of the resource.

c. The resilience of the resource in the face of change and impact.

The Sonoran-Mojave Creosotebush-White Bursage Desert Scrub and Sonoran Paloverde-Mixed Cacti Desert Scrub vegetation communities were identified as at risk on the basis of the regional trend analysis described in Section 2.1.5.2. Per the IDT analysis and stakeholder review of the criteria and analysis, the IDT identified the following residual impacts that may warrant compensatory mitigation:

Agua Caliente SEZ

- The loss of habitat and individuals of the Sonoran desert tortoise, a candidate species for listing under the Endangered Species Act (ESA) and the following BLM-sensitive animal species: Le Conte's Thrasher, California leaf-nosed bat, pale Townsend's big-eared bat, and western burrowing owl. Category 3 Sonoran desert tortoise habitat is located outside the SEZ to the north and northwest. Desert tortoises may still use lower quality habitat on the SEZ where they may be directly and indirectly impacted by solar development.

- The loss of ecosystem services and the human uses depending on them, as a result of development and until the lease expires and the site is restored. The primary components of an ecological system are: soils, vegetation communities, water, air, and wildlife.

Brenda SEZ

- The loss of habitat and individuals of the Sonoran desert tortoise, a candidate species for listing under the ESA and the following BLM-sensitive animal species: California leaf-nosed bat, pale Townsend's big-eared bat, and western burrowing owl. Category 2 desert tortoise habitat is located outside the SEZ to the south and west. Desert tortoises may still use lower quality habitat on the SEZ where they may be directly impacted by solar development.

- The loss of ecosystem services and the human uses depending on them, as a result of development and until the lease expires and the site is restored. The primary components of an ecological system are: soils, vegetation communities, water, air, and wildlife.

Gillespie SEZ

- The loss of habitat and individuals of the Sonoran desert tortoise, a candidate species for listing under the ESA, and the following BLM-sensitive animal species: Mexican rosy boa (suitable habitat not likely present in the recommended developable area of the SEZ), and California leaf-nosed bat, and western burrowing owl. Category 2 desert tortoise habitat is located outside the SEZ adjacent to the southern border. Desert tortoises may still use lower quality habitat on the SEZ where they may be directly impacted by solar development.

- The loss of ecosystem services and the human uses depending on them, as a result of development and until the lease expires and the site is restored. The primary components of an ecological system are: soils, vegetation communities, water, air, and wildlife.

Except where noted, for all three SEZs the following residual impacts were identified as having the potential to occur, depending on the way the area is developed, the success of avoidance and minimization, data gaps, and the discovery of unanticipated resources:

- Alterations to surface hydrology,

- Impacts on water quality and groundwater,

- Impacts on cultural resources are possible pending project-specific details and tribal consultation,

- Visual impacts on specially designated areas (Gillespie and Agua Caliente only),

- Impacts on visual resources,

- Environmental justice impacts (Agua Caliente and Brenda only),

- Certain Native American concerns (e.g., loss of habitat and cultural values),

- Impacts on recreation, and

- Impacts on livestock grazing, particularly loss of range improvements (Gillespie only).

2.5 Regional Goals and Mitigation Desired Outcomes

This strategy is focused on recommending appropriate compensation for the residual impacts of solar development in the Arizona SEZs that warrant mitigation (i.e., those impacts that cannot be either avoided or minimized onsite and are likely to exacerbate problematic regional trends) (Sec 2.4.3.2.2). For impacts recommended for regional compensatory mitigation, the mitigation desired outcome, at the narrowest level, is to offset the residual adverse impacts that are expected to occur onsite with actions that improve the impacted resource elsewhere in the region.

The Yuma, Lake Havasu, and Lower Sonoran RMPs (BLM 2010, BLM 2007, and BLM 2012b) guide BLM project-specific decisions in the region in which the Arizona SEZs are located. The RMPs establish management goals and guidance related to the residual impacts identified in Section 2.4.3.2.2 for the Arizona SEZs. The RMP guidance regarding regional goals and objectives is identified in the second column of Table 2-5.

The SEZ-specific desired mitigation outcomes and potential mitigation actions are presented in the third and fourth columns of Table 2-5. They are high-level desired outcomes to be considered in project-specific NEPA for selecting compensatory mitigation sites and actions within the region. Potential compensatory mitigation sites and actions for the Arizona SEZs are evaluated in Section 2.8.

Table 2-5. Summary table of regional goals, objectives, and mitigation desired outcomes and actions for the Arizona Solar Energy Zones

Resource Impacted that May Warrant Compensatory Mitigation	Regional Goals and Regional Objectives/RMP Guidance	Mitigation Desired Outcomes[8]	Potential Mitigation Actions
Ecosystem: Sonoran–Mojave Creosotebush–White Bursage Desert Scrub Vegetation Community	**Goal:** Preserve and/or restore creosote bursage and desert scrub ecosystem/habitat community disrupted by development (taking into account the existing landscape condition). **Objective:** Maintain or restore Sonoran-Mojave Creosotebush-White Bursage Desert Scrub in moderately-high to very-high intactness in the Sonoran Desert in Arizona. **RMP guidance:** Require mitigation where plants and parts of plants will be destroyed from a residual impact as a result of development, disturbance, or disposal. For BLM-authorized surface disturbing activities within desired plant communities, impacts to vegetation will be mitigated through: avoidance, minimization, soil stabilization and vegetative rehabilitation, transplanting appropriate species, salvage of plant and plant parts. (Yuma RMP) Identified as a "desired plant community" or "native plant/vegetative community." (Yuma, Lake Havasu, Lower Sonoran RMPs)	Create, restore, and/or acquire equivalent acreage of Sonoran-Mojave Creosotebush-White Bursage Desert Scrub lost through development on SEZ to 80% of existing vegetative cover (acres) and composition of primary plant species within 5 years of initiation of land disturbing development on the SEZ as an interim goal, with 100% as the end goal over 20 years. Create, restore, and/or acquire equivalent acreage of cryptogam cover lost through development on SEZ.	Habitat enhancement and acquisition. Close and revegetate unauthorized roads. Improve vehicle barriers and signage along WSA boundaries.
Ecosystem: Sonoran Paloverde–Mixed Cacti Desert Scrub Vegetation Community	**Goal:** Preserve and/or restore paloverde-mixed cacti ecosystem/habitat community disrupted by development (taking onto account the existing landscape condition). **Objective:** Maintain or restore Sonoran Paloverde-Mixed Cacti Desert Scrub in moderately-high to very-high intactness in the Sonoran Desert in Arizona. **RMP guidance:** Require mitigation where plants and parts of plants will be destroyed from a residual impact as a result of development, disturbance, or disposal. For BLM-authorized surface disturbing activities within desired plant communities, impacts to vegetation will be mitigated through: avoidance, minimization, soil stabilization and vegetative rehabilitation, transplanting appropriate species, salvage of plant and plant parts. (Yuma RMP) Identified as a "desired plant community" or "native plant/vegetative community." (Yuma, Lake Havasu, Lower Sonoran RMPs)	Create, restore, and/or acquire equivalent acreage of Paloverde-Mixed Cacti Desert Scrub ecosystem habitat community lost through development on SEZ to 80% of existing vegetative cover (acres) and composition of primary plant species within 5 years of initiation of land disturbing development on the SEZ as an interim goal, with 100% as the end goal over 20 years. Create, restore, and/or acquire equivalent acreage of cryptogam cover lost through development on SEZ.	Habitat enhancement and acquisition. Close and revegetate unauthorized roads.

---

8   The mitigation desired outcome is a measurable objective on the scale of an SEZ that is tied explicitly to the recommended mitigation action and can be applied to achieve the regional goals and objectives of the resource.

Table 2-5. (Cont.)

Resource Impacted that May Warrant Compensatory Mitigation	Regional Goals and Regional Objectives/RMP Guidance	Mitigation Desired Outcomes[9]	Potential Mitigation Actions
Riparian Vegetation	**Goal:** Preserve and/or restore riparian vegetation and habitat disrupted by development. **Objective:** Designation of non-development areas serves to avoid many impacts to this vegetation/habitat type. **Objective:** Maintain or restore riparian vegetation ecosystem habitat community in moderately-high to very-high intactness across 80% of its current distribution in the Sonoran Desert in Arizona. **RMP guidance:** Avoid desert wash woodlands to the greatest extent possible during BLM-authorized surface disturbing activities. (Yuma RMP) Identified as a "desired plant community" or "native plant/vegetative community." (Yuma, Lake Havasu, Lower Sonoran RMPs)	Create, restore, and/or acquire equivalent acreage of riparian vegetation ecosystem habitat community lost through development on SEZ to 80% of existing vegetative cover (acres) and composition of primary plant species within 5 years of initiation of land disturbing development on the SEZ as an interim goal, with 100% as the end goal over 20 years.	Maintain & Restore Wilderness Characteristics. Improve vehicle barriers and signage along WSA boundaries. Riparian system rehabilitation or restoration.
Special Status Species: Le Conte's Thrasher	**Goal:** Maintain viable populations (equal to or larger) of affected BLM or other special status species in the region. **SSS1: LeConte's thrasher** **Objective:** See objectives for creosote-bursage and desert scrub ecosystem and riparian vegetation and ecosystem Sonoran Paloverde-Mixed Cacti Desert Scrub Vegetation Community and habitat lost through development on the SEZ. To address impacts to LeConte's thrasher the pattern of vegetation established should be suitable to provide habitat. **RMP Guidance:** Per BLM Manual 6840, mitigation actions will seek to maintain, enhance, and restore SSS habitat.	Maintenance of equal (or greater) amount of habitat and pattern of vegetation.	Improve existing wildlife permeability habitat conditions: through restoration of old agricultural fields. Acquisition of habitat that meets standards for high quality habitat.

---

[9] The mitigation desired outcome is a measurable objective on the scale of an SEZ that is tied explicitly to the recommended mitigation action and can be applied to achieve the regional goals and objectives of the resource.

Table 2-5. (Cont.)

Resource Impacted that May Warrant Compensatory Mitigation	Regional Goals and Regional Objectives/RMP Guidance	Mitigation Desired Outcomes[9]	Potential Mitigation Actions
Special Status Species: California leaf nosed bat and Pale Townsend's big eared bat	**Goal:** Maintain viable populations (equal to or larger) of affected BLM or other special status species in the region. **SSS2: California leaf nosed bat and Pale Townsend's big eared bat** **Objective:** See objectives to for creosote- bursage and desert scrub ecosystem and riparian vegetation and habitat lost through development as offsetting impacts to vegetation should offset impacts to production of flying insects providing forage for these two bat species. **RMP Guidance:** Per BLM Manual 6840, mitigation actions will seek to maintain, enhance, and restore SSS habitat.	Maintenance of equal (or greater) amount of foraging habitat.	Improve existing wildlife permeability habitat conditions: Restore old agricultural fields. Habitat enhancement or habitat restoration.
Special Status Species: Sonoran Desert Tortoise	**Goal:** Maintain viable populations (equal to or larger) of affected BLM or other special status species in the region. **SSS3:** Desert Tortoise **Objective:** See objectives for creosote-bursage and desert scrub ecosystem and ecosystem Sonoran Paloverde-Mixed Cacti Desert Scrub Vegetation Community and habitat lost through development. To address impacts to Desert Tortoise the pattern of vegetation established should be suitable to provide habitat. **RMP Guidance:** No net loss of Category I and II desert tortoise habitat, in accordance with the Desert Tortoise Rangewide Plan and other applicable policy guidance. Enhance the conservation and management of desert tortoise habitat.	Maintenance of equal (or greater) amount of Sonoran desert tortoise habitat (if presence on the SEZ is identified during pre-construction survey). Restoration of wildlife connectivity.	Acquire non-federal lands that include tortoise habitat. Improve habitat quality of tortoise habitat outside of SEZ through reduction of stressors to offset loss of or reduction in quality of tortoise habitat on SEZ impacted through development. Restore wildlife connectivity and habitat conditions through restoration of old agricultural fields, modification of wildlife fencing to meet wildlife-friendly standards, construction of wildlife crossing structures (overpasses or underpasses) at key locations, and installation of tortoise fencing along key reaches. These connectivity measures would mitigate impacts from future development. Protect and promote tortoise movement between Buckeye Hills and North Maricopa Mountains through purchasing lands to preserve as linkages. This could include Highway SR85 design enhancements to reduce roadway mortality and/or improve permeability for tortoise.

Table 2-5. (Cont.)

Resource Impacted that May Warrant Compensatory Mitigation	Regional Goals and Regional Objectives/RMP Guidance	Mitigation Desired Outcomes[9]	Potential Mitigation Actions
Special Status Species: Western Burrowing Owl	**Goal:** Maintain viable populations (equal to or larger) of affected BLM or other special status species in the region. **SSS4: Western Burrowing Owl** **Objective:** See objectives for creosote-bursage and desert scrub ecosystem and ecosystem Sonoran Paloverde-Mixed Cacti Desert Scrub Vegetation Community and habitat lost through development. To address impacts to the Western Burrowing Owl the pattern of vegetation established should be suitable to provide foraging habitat. **RMP Guidance:** Per BLM Manual 6840, mitigation actions will seek to maintain, enhance, and restore SSS habitat.	Maintenance of equal (or greater) amount of nesting and foraging habitat. Establish a pattern of vegetation suitable to provide habitat.	Improve existing wildlife permeability habitat conditions through restoration of old agricultural fields. Acquire site(s) for relocation of owl nests disturbed by construction on the SEZ in advance so owls can be relocated prior to construction.
Cultural Resources[10]	**Goals:** Identify and preserve significant cultural resources and ensure that they are available for appropriate uses by present and future generations. Seek to reduce imminent threats to cultural resources and resolve potential conflicts from natural or human-caused deterioration or potential conflict with other resource uses by ensuring that all authorizations for land use and resource use will comply with the National Historic Preservation Act. **Objective:** Minimize potential development impacts on cultural resources through implementation of design features and mitigation measures including, but not limited to, proper siting and location and reduction of unnecessary site disturbance. **RMP Guidance:** Identify, preserve, and protect important cultural resources, and ensure that these resources are available for future generations. (Yuma, Lake Havasu, and Lower Sonoran RMPs). Reduce Threats, reduce or prevent damage, and resolve potential conflicts from naturally occurring or unauthorized human-caused damage or deterioration. (Yuma and Lower Sonoran RMPs).	Where possible, avoid cultural resources, particularly high concentrations, through identification of non-development areas within SEZs. Protect and preserve at-risk cultural resources to provide mitigation for residual impacts within 5 years of development of SEZ. Enhance present and future public use and enjoyment of cultural resources in the region to provide mitigation for residual impacts within 5 years of development of SEZ.	Implement security and enforcement measures. Implement interpretive and educational measures.

---

[10] Although during evaluation of residual impacts, cultural resources received a finding of "maybe" for having residual impacts warranting regional compensatory mitigation (100% surveys of each of the SEZs indicate, despite avoidance of much of the areas containing cultural resources, compensation may be warranted for cultural stresses on the landscape), it is included in this table to aid in future discussions regarding development.

Table 2-5. (Cont.)

Resource Impacted that May Warrant Compensatory Mitigation	Regional Goals and Regional Objectives/RMP Guidance	Mitigation Desired Outcomes[9]	Potential Mitigation Actions
Visual Resources	**Goal:** Preserve and/or enhance scenic quality in the region through preservation of open-space landscapes and undisturbed views, or through restoration of habitat to compensate for visual resources impaired by development. **Objective:** Minimize potential development impacts on visual resources through implementation of design features and mitigation measures including, but not limited to, proper siting and location, color treatment, and reduction of unnecessary site disturbance. For example, cultural and ecological mitigation actions could in concert appropriately mitigate visual resource impacts through re-vegetation, increased site protection, etc. Visual resource mitigation efforts could also benefit cultural and ecological resources. **RMP guidance:** Restoration projects will ensure that visual resource impacts are minimized in the short term (5 years) and that Visual Resource Management (VRM) objectives in the project area are met in the long-term (life of the project). (Lower Sonoran RMP)	Repair, maintain, and/or enhance scenic quality by reducing visual contrast through proper landscape remediation and reclamation that restore natural scenic quality and integrity.	Implement proper landscape remediation and reclamation on derelict landscapes such as decommissioned construction sites, abandoned roads, etc. through restorative landform grading, soil treatment, and revegetation that results in the reduction of visual contrast. Reduce visual contrast of existing cultural modifications through color treatment and/or vegetative screening that reduce visual contrast and enhance overall landscape scenic quality.

2.6 Calculating the Recommended Mitigation Obligation for Arizona Solar Energy Zones

This section provides the BLM-recommended compensatory mitigation fees for the Arizona SEZs based on the residual impacts identified in Section 2.4.3.2. In Arizona, two options exist to satisfy compensatory mitigation obligations for residual impacts of solar development in SEZs after avoidance and minimization measures have been applied: proponent-responsible compensatory mitigation and contributions to a compensatory mitigation fund.

For contributions to a mitigation fund, the long-term responsibility for compensatory mitigation would be transferred away from the authorized land user (developer) to the fund manager, with payment of a predetermined fee based on the type and magnitude of the identified residual impacts warranting compensatory mitigation. The entire fee would be paid at the time development commences, but would be managed to provide for the selected mitigation actions over the life of the solar project impacts. If contribution to a mitigation fund is selected as the mitigation method in coordination with the developer, the likely fee for each of the Arizona SEZs will be identified before parcels are made available for auction. The fee will include updates to reflect current costs of acquisition and/or restoration, and may also include costs for compensatory mitigation for impacts warranting mitigation not previously included in the fee (e.g., cultural impacts and Native American concerns). Also, just prior to issuing a notice to proceed with construction, BLM may adjust that fee in order to include costs based on impacts that require consideration of project-specific data (e.g., impacts on visual resources). The final compensatory mitigation fee will be paid by the developer at the issuance of the Notice to Proceed (see Table 1-1).

The recommended compensatory mitigation fees for the Arizona SEZs were calculated based on the method described in the draft Procedural Guidance (BLM2014a), somewhat modified based on stakeholder input. The recommended compensatory mitigation fee is presented separately for the three AZ SEZs based on differences in anticipated impacts and associated costs for compensatory mitigation actions.

The specific values used to calculate the per-acre compensatory mitigation fee can vary between SEZs and involve a number of different calculations. Figure 2-11 presents a flow diagram describing the various potential pathways that can be used to calculate the per-acre regional compensatory mitigation fee. The steps that follow correspond to Figure 2-11 and outline the calculation of a recommended compensatory mitigation fee.

Step 1: Identify the mitigation technique for the obligation: The fee for Arizona will be based on a combination of acquisition and restoration.

Step 2: Estimate the costs and calculate the base fee: The market analysis for the Arizona SEZ mitigation consisted of a BLM biologist querying local contractors, range specialists, and realty staff for the cost of acquiring and restoring an acre of relevant vegetation for each SEZ. A BLM biologist determined the number of trees impacted for each SEZ based on a sample survey, this number is used in the calculation of the base fee. Similarly, the BLM biologist used aerial imagery data to estimate that vegetative cover on each SEZ is approximately 25%. Therefore, replacement of vegetation equal to approximately 25% of the SEZ area would be required.

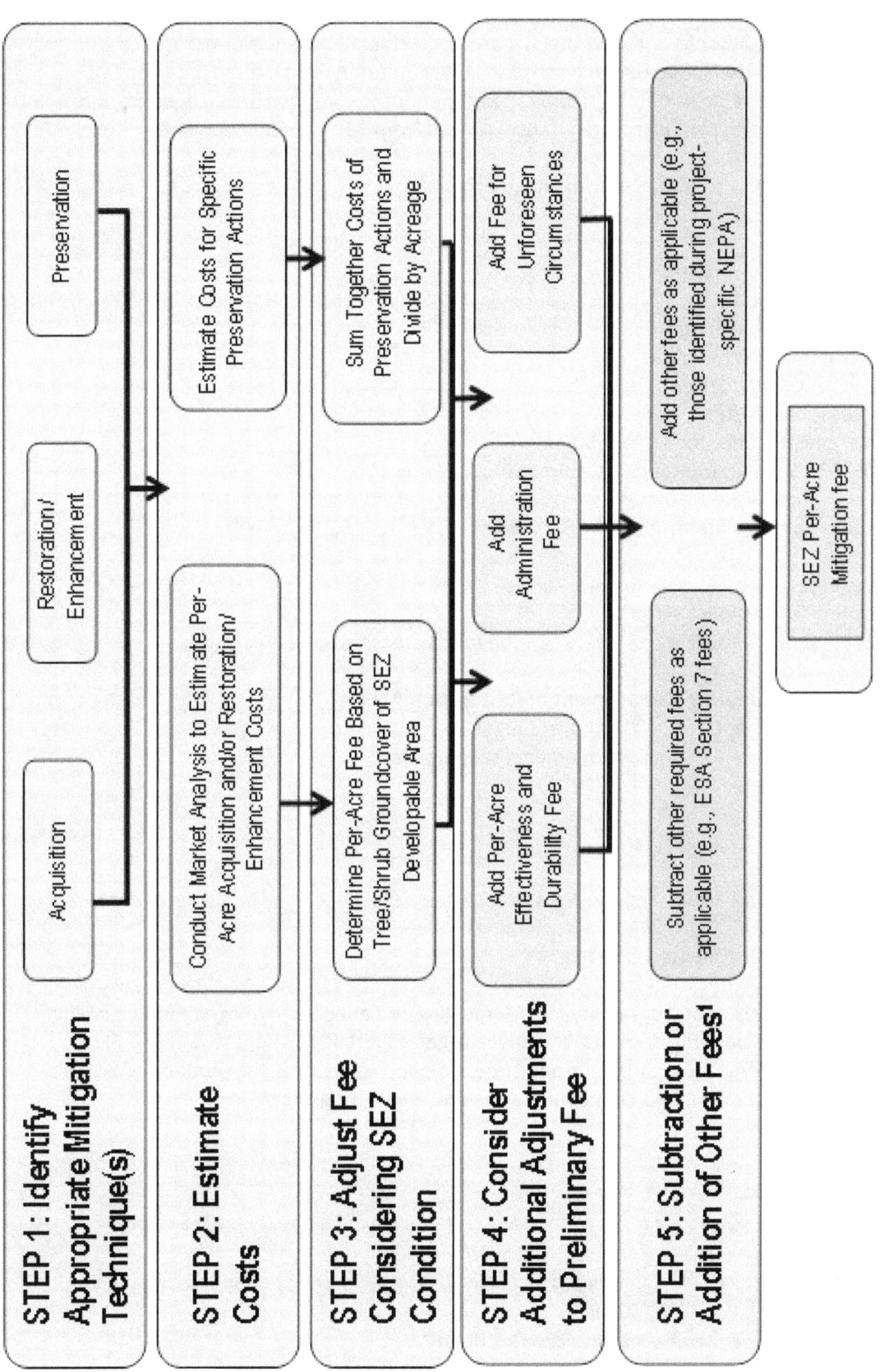

Figure 2-11. Steps for calculating per-acre regional compensatory mitigation fees based on impacts

[1] No ESA Section 7 fees are currently applicable for the Arizona SEZs. Any additional fees identified during project-specific NEPA will be added to the final mitigation fee as part of the NEPA decision.

Acquisition

Land acquisition for restoration site(s), based on actual BLM acquisitions, of undeveloped lands, in the Yuma, Lake Havasu, and Lower Sonoran Field Offices between 2012 and 2014. Fair market value as determined by Department of the Interior (DOI) Office of Valuation Services (which BLM is required to use) ranged from $325 to $500 per acre.)

Total Acquisition = $500/acre x impacted acres

Restoration

Tall Pot Method
Plant material $75/tree
DriWater System $75/tree
Browser Cage $15/tree
Installation $20/tree
Plant maintenance for 1 year $50/tree
10% $23.50 to account for soft estimate
        Total = $258.50/tree
Number of trees (determined by BLM biologist) x 3 at $258.50/tree[11]

Seeding
Seeding Open Desert Scrub[12]:
$1800/acre x 2 applications x 25%[13] of the total acres impacted

Short-term Adaptive Management of Restoration Actions[14]:
Tree Replacement: 100% of trees planted
Reseed: 100% of original seeding effort

Step 3: Calculate the adjusted base fee:

Adjusted base fee = land acquisition fee + restoration fee

Step 4: Consider additional adjustments to the fee:

Step 4A: Calculate an effectiveness and durability fee: Monitoring and adaptive management, beginning at project implementation, will identify any need to replace vegetation that does not thrive. The BLM recommends that the effectiveness and durability adjustment be applicable over the duration of project impacts; 50 years is assumed for mitigation implementation and monitoring.

---

[11] Three trees are planted for every one tree removed to ensure replacement of foliage volume in a timely manner.

[12] The seed mix used will be based on field sampling and evaluation of vegetation composition and cover on the portion of the SEZ to be impacted by development.

[13] Percentage is based on 25% desert scrub vegetative cover on the SEZ.

[14] Costs of monitoring are addressed in Step 4a.

The cost for long-term monitoring of the success of restoration is currently estimated to be $5 per acre per year, as used in the Dry Lake SEZ SRMS (BLM 2014b). This assessment assumed the annual monitoring cost of $5/acre over the duration of 50 years (that is, $250 per acre total).[15]

Adjusted base fee = Step 3 adjusted base fee + $250/acre

Step 4B: Include an administration fee of 5% to be used for management and reporting of regional compensatory mitigation funds and a fee to account for any unforeseen future circumstances. A 10% unforeseen future circumstances fee has been selected based on the Lower Colorado River Multi-Species Conservation Program (LCRMSCP 2004) and professional judgment. This fee will allow BLM to monitor and respond to episodic events that may not be evaluated in a timely manner during routine long-term monitoring as discussed in Step 4A. These episodic events (e.g., fire or flood) may be more likely to result in the need for larger, more involved corrective actions beyond just reseeding and replanting in small localized areas. If at the end of the 50 year period, portions of the unforeseen circumstances fee remain unused, the money will be pooled for future mitigation efforts requiring additional contingency monies. An administrative fee of 5% plus an unforeseen circumstance fee of 10% equals a 15% adjustment:

Adjusted Fee = Step 4A adjusted base fee x 1.15

Step 5: Subtract or add other fees: Add other fees as applicable (e.g., those identified during project-specific NEPA). The Arizona SEZs are not located in an area subject to any Section 7 permitting fees for federally-listed species under the ESA. Because there are currently no ESA-listed species expected to be affected by solar energy development on the SEZs, no fee adjustment is currently anticipated; however, additional costs or reductions may be identified on the basis of the impact evaluation during project-specific NEPA.

At this time, the recommended mitigation fee does not include a component for mitigation of cultural resources and some visual resources, because consultation for cultural resources has not occurred at the project level, and because of the project-specific nature of impacts on visual resources. If compensatory mitigation is identified as warranted for addressing cultural resource and/or visual resource impacts during future project-specific evaluations, some additional cost may be determined separately from the process described in this section. Compensatory mitigation also may be considered to address any residual socioeconomic and/or environmental justice impacts, if identified prior to project approval. This could be included in the mitigation fee or take the form of direct contributions from the developer to a community fund or in-kind contribution to affected local governments or populations.

Recommended SEZ per acre mitigation fee =
(Step 3 adjusted base fee + $100,000) x 1.15)/number of acres impacted

Table 2-6 provides the estimated number of trees impacted and estimated restoration costs for each of the Arizona SEZs. In the Draft SRMS, the preliminary per acre fee ranges depended on the degree of replacement through adaptive management; the low end of the range assuming 50%

---

[15] Factors which may be considered in adjusting the fee collected for monitoring at the time of project-specific NEPA include but are not limited to monitoring methodology and frequency, opportunities to utilize remote sensing, distance to the mitigation site, and travel costs (e.g., if overnight accommodations needed).

replacement and the high end 100% replacement. The Agua Caliente SEZ fee ranged from $2,949–$3,712 per acre, the Brenda SEZ ranged from $3,140–$3,964 per acre, and the Gillespie SEZ ranged from $3,436–$4,368 per acre. However, based on stakeholder comments and consideration of planting success, BLM recommends a mitigation fee for each SEZ based on 100% replacement. The likely compensatory mitigation fee for each SEZ will be identified as part of the pre-auction NEPA decision record, and may include adjustments for land value and inflation and costs for impacts not previously included (e.g., for cultural resource or visual resource impacts).

Table 2-6. Components of the recommended per acre compensatory mitigation fees for the Arizona Solar Energy Zones

	Agua Caliente SEZ	Brenda SEZ	Gillespie SEZ
Developable Acres	2,021 acres	1,906 acres	2,213 acres
Number trees[16]	1,112	1,298	2,035
STEP 2:			
Acquisition Cost ($500 * developable acres)	$1,010,500	$953,000	$1,106,500
Restoration: Tall Pot Method # trees * (3*$258.5)	$862,356	$1,006,599	$1,578,143
Restoration: Seeding $1,800*2 applications * (25% of total acres)	$1,818,900	$1,727,200	$1,991,700
Adaptive Management:			
Tree replacement: 100% (2:1 mitigation ratio)	$862,356	$1,006,599	$1,578,143
Reseed: 100% effort (2:1 mitigation ratio)	$1,818,900	$1,727,200	$1,991,700
STEP 3:			
SEZ Base Fee	**$6,373,012**	**$6,420,598**	**$8,246,186**
STEP 4:			
Effectiveness & Durability Fee	$505,250	$476,500	$553,250
Adjusted Base Fee Subtotal	$6,878,262	$6,897,098	$8,799,436
Administration Fee	5%	5%	5%
Unforeseen Future Circumstances Fee	10%	10%	10%
Adjusted Base Fee Subtotal	$7,910,001	$7,931,663	$10,119,351
STEP 5:			
Other Fees (ESA, etc.) – none currently identified	0	0	0
Adjusted Base Fee	$7,910,001	$7,931,663	$10,119,351
Per Acre Fee	**$3,914**	**$4,161**	**$4,573**

Prior to collecting the fee and after the project-specific NEPA evaluation, it may again be adjusted for inflation and/or for costs not previously included (e.g., for cultural resource or visual resource impacts). The BLM recommends the following per-acre compensatory mitigation fees (2015 dollars) of $3,914 per acre for Agua Caliente SEZ, $4,161 per acre for Brenda SEZ, and $4,573 per acre for Gillespie SEZ.

---

[16] Ironwood (*Olneya tesota*), Mesquite (*Prosopis* spp.), and Paloverde (*Cercidium* spp.) on developable area (based on sample)

2.7 Management of Solar Regional Compensatory Mitigation Obligations

The BLM will select management options for SEZ mitigation obligations that are consistent with the BLM's interim regional mitigation policy, draft Manual Section 1794, issued June 13, 2013 and DOI's Departmental Manual Part 600 DM 6 Landscape-Scale Mitigation Policy (DOI 2015), issued October 23, 2015, which include guidance for management of funds collected as part of the restoration, acquisition, or preservation portion of the total mitigation obligation by an independent third party. BLM Arizona will incorporate the most recent departmental mitigation policy to implement a transparent and effective accounting system to track funds contributed and funds spent, and to establish a funding mechanism to cover administration, durability, monitoring, and reporting for the investments for the duration of the impacts from development in the SEZs.

As a possible example of third party management of funds, the National Fish & Wildlife Foundation (NFWF) is a private non-profit organization charted by Congress with the ability to work nationally and across state and local political boundaries. They are a well-established and transparent financial management service. As a neutral third party NFWF offers low management fees and tax-free growth of funds resulting in more money for on-the-ground conservation. NFWF ensures efficient contracting and administration; there is no risk of funds being diverted to government treasuries or other uses. The administration fee is intended to be used towards a 3rd party's management fees, such as those for NFWF.

2.8 Evaluation of Compensatory Mitigation Sites, Actions, and Desired Outcomes

The proposed regional compensatory mitigation sites and actions will mitigate for the loss of some of the resources that will occur as a result of solar development in the Arizona SEZs. The BLM Arizona State Office considered several regional compensatory mitigation action alternatives. The suite of potential mitigation actions were generated by soliciting proposals from the public and from BLM staff. The proposals included:

1. Restoring disturbed land in several ACECs in the vicinity of the Arizona SEZs.
   a. Restoring desired vegetation in areas where the natural vegetative regime has been altered
   b. Eradicating invasive species
   c. Fencing
   d. Closing and restoring or revegetating unauthorized roads

2. Acquiring private land through purchase or easement, with appropriate resources, in the vicinity of the Arizona SEZs and managing it for conservation values.
   a. Seeking to acquire non-federal lands within or adjacent to lands within an ACEC

3. Maintaining and restoring wilderness characteristics.

4. Closing and revegetating unauthorized roads within the Cactus Plain Wilderness Study Area (WSA) boundaries.

5. Securing Sonoran desert tortoise habitat through acquisition of an amount at least equal to that converted to non-habitat through development on the SEZs.

6. Improving habitat quality of tortoise habitat outside of the SEZs through reduction of stressors to offset loss of or reduction in quality of tortoise habitat on the SEZs impacted through development.

7. Securing site(s) for relocation of burrowing owl nests disturbed by construction on the SEZs in advance so owls can be relocated prior to construction.

8. Acquiring non-federal lands to consolidate management and to establish a broader corridor of conservation management across Rainbow Valley.

9. Implementing security and enforcement measures.

10. Implementing interpretive and educational measures for cultural resources.

11. Strengthening the management prescriptions for the area in a future RMP amendment.

12. Establishing wildlife crossings to maintain connectivity.

The following proposed mitigation sites were given a preliminary score using criteria based on the regional compensatory mitigation goals described in Section 2.5. The results of the scoring are summarized in the matrix table for candidate regional compensatory mitigation sites for the Arizona SEZs (Appendix D):

- Cactus Plain WSA
- Hoodoo Wash
- Palomas Plain
- Cactus Plain
- La Posa Plain
- Lower Gila River Terraces Package
- Quail Point
- Boa Sorte
- Cocoraque Butte
- La Osa Ranch
- Ranegras Plain
- Sacaton Flats
- Saddle Mountain ACEC
- Rainbow Valley
- Sears Point ACEC
- Fred J. Weiler Vegetation Habitat Management Area
- Los Robles Archaeological/Historic District
- Marana Mound
- Ajo

The following criteria were used to rank these sites relative to their values and ability to mitigate the residual impacts identified:

- Site and its proposed actions would mitigate for all or most identified residual impacts that warrant compensatory mitigation.

- Site and its proposed actions meet regional conservation/mitigation goals, objectives, and desired outcomes.

- The site is within the area of impact within Lake Havasu, Yuma, or Lower Sonoran Planning Areas (i.e., the same subregion and landscape context as the Arizona SEZs).

- The site contains the same dominant vegetation community.

- The site provides habitat for a similar suite of general wildlife, special status wildlife, and rare plants.

- The degree to which the compensatory mitigation site and actions are consistent with the Lake Havasu, Yuma, or Lower Sonoran RMPs.

- The degree to which applicable management prescriptions in the Lake Havasu, Yuma, or Lower Sonoran RMPs facilitate durable mitigation investments. Management prescriptions that facilitate durability include, but are not limited to: special conservation-oriented designations, such as national conservation areas, ACECs, designated wilderness areas, and wilderness study areas; areas where land-disturbing activities are prohibited; and areas where land-disturbing activities are discouraged.

- Compensatory mitigation at the site would be feasible (as indicated by level of documentation, difficulty of implementation, time frame needed to establish the site and achieve mitigation goals and objectives, and the cost estimate for the compensatory mitigation actions).

- Effectiveness of the action based on an assessment of how effective the mitigation will be in the context of achieving mitigation goals and objectives.

- That the mitigation will consist of actions that would not otherwise be undertaken by the BLM (additionality).

- Risk of failure of compensatory mitigation actions at the site, based on known constraints and known current and future surrounding land uses.

Proposed Mitigation Actions and Sites

The size of the Arizona SEZs could allow for the siting of multiple utility-scale solar energy projects in a single SEZ at different times over the course of SEZ development. The technology, scale, and schedule of these developments would influence the prioritization of compensatory mitigation options. For this reason, the BLM is currently considering many of the potential mitigation actions and sites listed above and in Appendix D. Appendix D includes the matrix and presents all of the candidate site locations in relation to the SEZs in Figures D-1 and D-2. The determination of required compensatory mitigation actions and sites will be conducted at the project level through a project-specific NEPA assessment, which would tier to the Solar PEIS and the Final RDEP EIS and consider recommendations from this SRMS document.

Figure 2-12 displays the locations of the highest scoring regional compensatory mitigation sites for the Agua Caliente and Gillespie SEZs based on stakeholder and BLM input in the candidate site matrix (Appendix D), namely Sears Point ACEC, Rainbow Valley, the Lower Gila River Terraces Package, and Quail Point. The top scoring site, Sears Point ACEC, was nominated separately by three different groups (The Nature Conservancy, BLM Yuma Field Office, and Archaeology Southwest) for its creosote-white

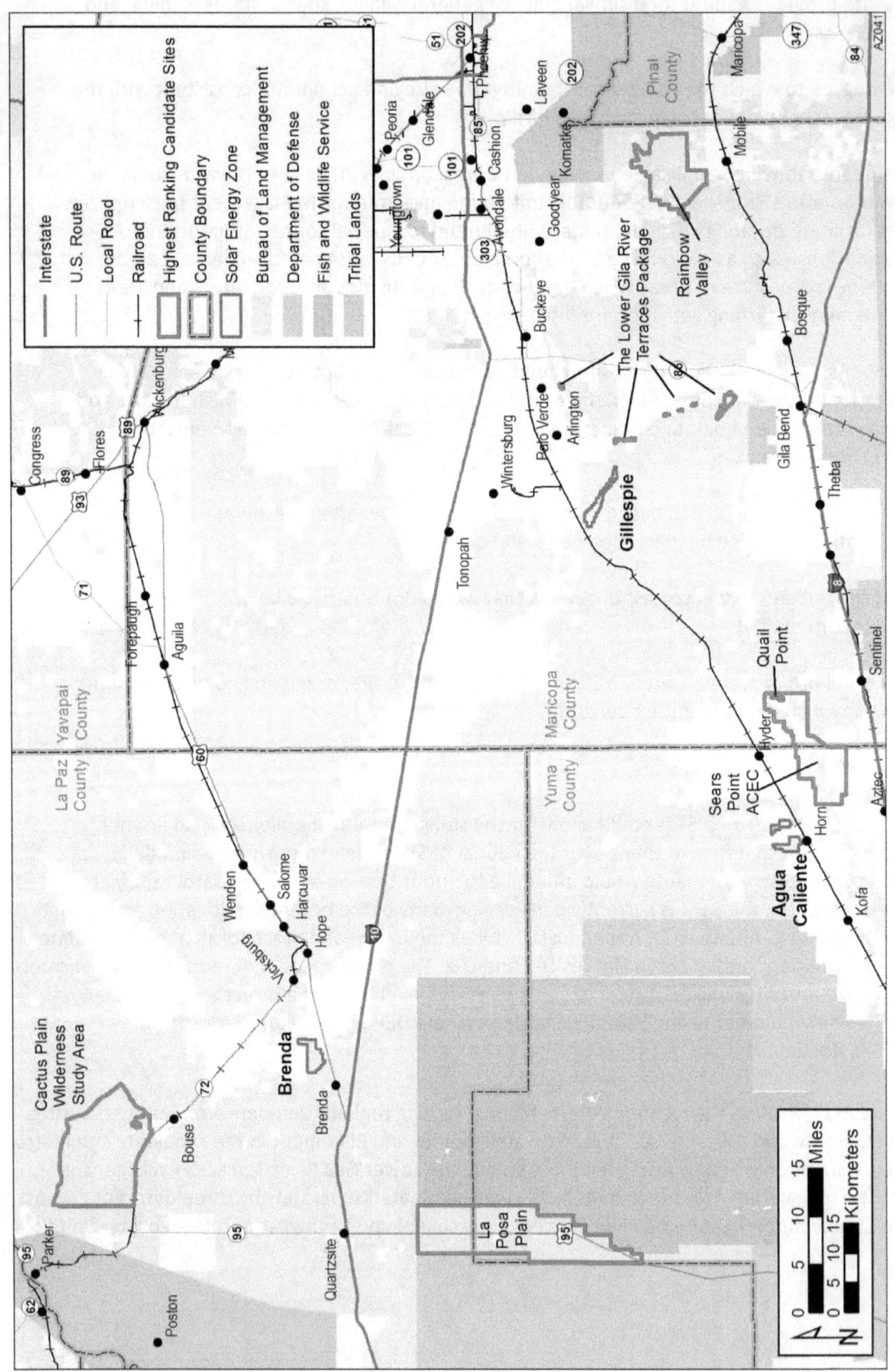

Figure 2-12. Highest scoring candidate site locations based on the Candidate Site Matrix, Appendix D

bursage habitat and cultural resources. Recommended mitigation actions are acquisition and restoration which could preserve and increase intactness and connectivity of landscapes and habitat, as well as protect BLM sensitive species and cultural resources. Proposed actions, such as road closure and revegetation and restoration of agriculture fields, would help meet regional conservation/mitigation goals and objectives. The site scored high in the categories of feasibility, effectiveness, additionality, risk and durability. Bonus points were added to the preliminary score based on the presence of BLM sensitive species, desert washes, riparian areas, and significant cultural resources.

Rainbow Valley, recommended by The Nature Conservancy (TNC) for its habitat values and presence of BLM special status species, scored well. Acquisition and restoration, including the removal of barriers and creating wildlife crossing structures, are recommended as mitigation actions. These actions would protect tortoise habitat, preserve and restore creosote/bursage habitat, and protect BLM sensitive species and cultural resources. The site received the maximum bonus points for additional criteria based on the presence of unique and/or valuable resources at the site and the linkage of the Sonoran Desert National Monument and the Sierra Estrella Mountains.

Quail Point and the Lower Gila River Terraces sites were nominated by Archaeology Southwest for their cultural resources including rock art, petroglyphs, and Archaic, Hohokam, and Patayan archaeology. Recommended mitigation actions are acquisition, habitat restoration, and access control which could preserve and increase intactness and connectivity of landscapes and habitat, as well as protect BLM sensitive species and cultural resources. Proposed actions would help meet regional conservation and mitigation goals and objectives, including protection of the cultural resources found at these sites.

The highest scoring mitigation sites for the Brenda SEZ (Figure 2-12) were Cactus Plain WSA and La Posa Plain, recommended by TNC. Acquisition and restoration actions were identified to protect the creosote-white bursage habitat and BLM sensitive species present at the sites. Proposed actions include closure and revegetation of unauthorized roads, removal of barriers, wildlife crossing structures, and tortoise fencing to meet regional mitigations goals.

2.9 Mitigation Effectiveness Monitoring and Adaptive Management Plan

In the Solar PEIS ROD, the BLM committed to developing and incorporating a monitoring and adaptive management plan into its solar energy program. The BLM "Assessment, Inventory, and Monitoring Strategy for Integrated Renewable Resources Management" (AIM Strategy) (Toevs et al. 2011) will guide the development of an Arizona Solar Energy Zone monitoring plan that will inform management questions at multiple scales of inquiry (e.g., the region/landscape, mitigation area, and project area). Detailed information about how the AIM Strategy will be implemented to support long-term monitoring of solar development is provided in Appendix A, Section A.2.4 of the Final Solar PEIS. This monitoring plan will also be consistent with and complement the BLM regional and national monitoring activities.

In the context of solar energy development, long-term monitoring should be conducted to (1) evaluate the effectiveness of mitigation measures, including avoidance, minimization, and regional compensatory mitigation; (2) detect unanticipated direct and cumulative impacts at the project and regional level; and (3) evaluate the effectiveness of elements of the BLM's solar energy program (e.g., policies, design features). To ensure that investments in regional compensatory mitigation actions are effective and that regional compensatory mitigation goals and outcomes are being met, it is critical that the long-term monitoring plan include monitoring outcomes specific to the regional compensatory mitigation sites and actions. The findings of the long-term monitoring activities will be examined by the BLM to support adaptive management of solar development (i.e., to identify the need to adjust operational parameters, modify mitigation measures, and/or implement new mitigation to prevent or minimize further impacts). The following steps will be conducted to develop the mitigation effectiveness monitoring plan for the Arizona SEZs:

Step 1. Develop Management Questions and Monitoring Goals

The BLM IDT has developed management questions to articulate the issues of concern related to monitoring mitigation effectiveness. The management questions provide the basis for developing monitoring goals. The management questions and monitoring goals for the Arizona SEZs are provided in the two text boxes that follow.

Management Questions Established for the Arizona Solar Regional Mitigation Strategy

- Are the design features of the solar development effectively containing impacts of solar installation to the project site (e.g., trend of attributes, special status species habitat indicators, invasive species, habitat metrics)?
- Are the avoidance areas maintaining ecological composition and process similar to those adjacent to the project area?
- Are the avoidance areas for cultural resources sufficient to protect their values from unintended or unanticipated adverse effects?
- Are the regional compensatory mitigation actions achieving their outcomes?
- Are the Arizona Solar Energy Zones (SEZs) mitigation actions collectively effective in improving the trend of landscape health metrics in the regional enhancement(s)?
- What are the status and trend of landscape health metrics for critical ecological processes necessary to sustain the Sonoran Desert ecosystem at two scales: the Arizona SEZs 2-mile buffer area and the compensatory mitigation area(s)? Note: Some impacts may need to be assessed at different distances (e.g., watershed, airshed).

Monitoring Goals Established for the Arizona Solar Regional Mitigation Strategy

1. Establish baseline measurements of landscape patterns and health. (Contributes to answer to Management Question (MQ) 1, 2, 4, and 5)
2. Establish baseline measurements for cultural resources values and determine the status and trend of these values once the permitted activity and related mitigation actions have been implemented. (Contributes to answer to MQ 1, 3, and 4)
3. Determine the status, condition, and trend of priority resources and landscape health metrics once the permitted activity and related mitigation actions have been implemented. (Contributes to answer to MQ 5)
4. Leverage the quantitative data from goals 1, 2 and 3 to map the location, amount, and spatial pattern of priority resources and disturbances. (Contributes to answer to all MQs)
5. Generate quantitative and spatial data to address goals 1 and 3 and to contribute to existing land health assessment and evaluation processes at multiple scales of inquiry. (Contributes to answer to MQ 6)
6. Generate quantitative and spatial data to determine if management actions (e.g., stipulations, land treatments) are moving resources toward desired states, conditions, or specific resource objectives identified in planning or related documents or legal mandates. (Contributes to answer to all MQs)
7. Use the collected data to validate and refine the conceptual understanding of key ecosystem components, processes, and sustainability concepts for the Sonoran Desert ecoregion and the Arizona SEZs. (Contributes to answer to MQ 6)

Step 2. Identify Measureable Monitoring Outcomes and Indicators

Measureable monitoring outcomes will be established for each monitoring goal identified in Step 1. Outcome setting will be based on current regulatory requirements, RMP goals, or the desired future condition consistent with the land potential (as described in the ecological site description, if available – see Step 4). Examples of measureable monitoring outcomes are provided in the text box titled Measureable Monitoring Outcome Examples.

Measureable Monitoring Outcome Examples

Examples of a measureable outcome for land status/trend of vegetation are:

Detect a difference of 10 percentage points in the average amount of bare ground in the <MITIGATION SITE> over a 5-year period with 80% confidence.

Determine whether at least 25% perennial grass cover in the <MITIGATION SITE> has been maintained with 90% confidence.

An example of an outcome for cultural resource values is:

Detect any unanticipated impacts attributable to development-related changes in natural processes (e.g., erosion, vegetation growth or removal) or human effects (e.g., trampling, casual collection, vandalism) associated with increased project-related access.

Outcome setting includes specifying the attribute and measurable indicators of those attributes to be monitored. Monitoring outcomes will indicate the allowable amount of change (specific) and confidence level for the measured change (measurable), relationship to the management question (relevant), and time frame during which the measurement occurs to effectively inform management (time sensitive).

Indicator selection will start with the standard AIM core and contingent quantitative indicators (MacKinnon et al. 2011) and supplement with additional indicators derived from ecosystem conceptual models and/or linked to specific management questions. The AIM core indicators and methods provide high-quality quantitative information on all land cover types the BLM manages (MacKinnon et al. 2011). Table 2-7 (reproduced from MacKinnon et al. 2011) lists each method and the corresponding indicators it measures, and the table describes recommendations to achieve consistent implementation across the BLM. When an ecological site at a monitoring site is identified, the BLM core measurements can be assessed in concert with information contained in the ecological site descriptions and the accompanying state and transition model to ascertain departure from an expected reference condition. The methodology for this assessment is contained in "Interpreting Indicators of Rangeland Health," BLM Technical Reference 1734-6. Table 2-8 is a summary table from this technical reference.[17]

In addition to the BLM core indicators, the design features for the Solar PEIS indicate that the BLM will consider requiring dust and noise monitoring as a leasing stipulation for the Arizona SEZs (BLM 2012). The developer's proposal will be reviewed by the BLM monitoring team to evaluate the

---

[11] Tables 2-7 and 2-8 summarize guidance for BLM monitoring that may change over time; the most current versions of these guidance documents should be used at the time the monitoring program for the SEZs in Arizona is established.

Table 2-7. Recommended methods and measurements for core and contingent indicators (reproduced from MacKinnon et al. (2011))

Method	Indicator(s)	Description
For core indicators		
Line-point intercept with plot-level species inventory	• Bare ground • Vegetation composition • Nonnative invasive species • Plant species of management concern	Line-point intercept (LPI) is a rapid and accurate method for quantifying cover of vegetation and bare ground. Because LPI can underestimate cover of uncommon species, this method is supplemented with searches of a 150-ft (45.7-m) diameter standard plot for at least 15 minutes and until new species detections are more than 2 minutes apart. When performing LPI within tree cover, a modified pin method (e.g., a pivot-table laser or extendable pin) will be used to capture overstory cover.
Vegetation height measurement	• Vegetation height	Measure height of tallest leaf or stem of woody and herbaceous vegetation (living or dead) within a 6-in (15-cm) radius recorded for points along a transect. If vegetation is taller than 10 ft, a standard tape and clinometer method should be used to estimate vegetation height.
Canopy gap intercept	• Proportion of soil surface in large intercanopy gaps	Canopy gap intercept measures the proportion of a line covered by large gaps between plant canopies and is an important indicator of the potential for erosion. Use 1-ft (30-cm) minimum gaps.
For contingent indicators		
Soil stability test	• Soil aggregate stability	This test measures the soil's stability when exposed to rapid wetting and provides information on integrity of soil aggregates, degree of structural development, resistance to erosion, and soil biotic integrity.
Soil sample collection and analysis	• Significant accumulation of soil toxins	The presence and concentrations of toxins are assessed by collecting three samples from the soil surface and one sample at depths of 0 to 4 in (0 to 10 cm) and 4 to 8 in (10 to 20 cm) using a soil corer and following Forest Inventory and Analysis protocol.

efficacy of the proposal in complying with permit stipulations and informing BLM regulatory and land management needs.

Special Status Plant Species Monitoring. At this time there are no known occurrences of special status plant species on the Arizona SEZs. The BLM will consider requiring the developer to conduct long-term monitoring on special status plant populations if found on the project site and located in the same geographic region for the length of the duration of the impact. If applicable, a special status plant species monitoring plan will be designed to determine the status, trend, and recruitment success of the populations and will follow methods described in BLM Technical Reference 1730-1, "Measuring and Monitoring Plant Populations" (Elzinga, Salzer, and Willoughby 1998).

Table 2-8. Quantitative indicators and measurements relevant to each of the three land health attributes (reproduced from Pellant et al. (2005))

Attribute	Qualitative Assessment Indicator	Quantitative Measurement Method	Key Quantitative Assessment Indicator
Soil/site stability	• Rills • Water flow patterns • Pedestals and/or terracettes • Bare ground • Gullies • Wind-scoured, blowout, and/or depositional areas • Litter movement • Soil surface resistance to erosion • Soil surface loss or degradation • Compaction layer	Line-point intercept	Bare ground
		Canopy gap intercept	Proportion of soil surface covered by canopy gaps longer than a defined minimum
		Soil stability test	Soil macro-aggregate stability in water
Hydrologic function	• Rills • Water flow patterns • Pedestals and/or terracettes • Bare ground • Gullies • Soil surface resistance to erosion • Soil surface loss or degradation • Plant community composition and distribution relative to infiltration and runoff • Compaction layer • Litter amount	Line-point intercept	Bare ground
		Canopy gap intercept	Proportion of soil surface covered by canopy gaps longer than a defined minimum
		Soil stability test	Soil macro-aggregate stability in water
Biotic integrity	• Soil surface resistance to erosion • Soil surface loss or degradation • Compaction layer • Functional/structural groups • Plant mortality/decadence • Litter amount • Annual production • Invasive plants • Reproductive capability of perennial plants	Soil stability test	Soil macro-aggregate stability in water
		Line-point intercept	Plant canopy (foliar) cover by functional group
		Line-point intercept	Plant basal cover by functional group
		Line-point intercept	Litter cover
		Line-point intercept	Invasive plant cover

Step 3. Develop Sampling Schema

Based on the management questions, monitoring goals, measurable outcomes, and the indicators developed in Steps 1 and 2, the BLM IDT will determine the temporal and spatial scale of data collection activities. To develop the sampling schema, the following work will be conducted:

Develop a Statistically Valid and Scalable Sampling Design. Ecological sites are areas of land with the potential to produce similar types and amounts of vegetation based on soils and climate, and are the basic units for stratifying landscapes for BLM monitoring activities. Because ecological site descriptions describe the ecological states (plant communities) that can occur within the ecological site and can provide expected indicator values for reference states, they are the foundation upon which BLM monitoring data are evaluated. These data are also fundamental for terrestrial upland land health standards and land health evaluations. Where ecological site descriptions have not been developed, land potential metrics can be developed using a combination of field and remote sensing data to describe current and potential future conditions at broad scales.

Incorporate Status and Trend Monitoring. The monitoring locations are determined through a statistically based (i.e., randomized) selection of monitoring sites. Once the monitoring extent (i.e., inference area) is determined for each scale, a stratified random technique will be used to select monitoring sites such that every location within the monitoring extent has a known and nonzero probability of being selected for sampling. Strata will be based on ecological sites (or groupings of sites with similar ecological characteristics) to allow for adequate representation of ecological characteristics and linear features (e.g., ephemeral washes). Locations would be monitored in a manner consistent with the BLM's AIM Strategy (Toevs et al. 2011) in order to understand status and trends in monitored resources. This example sampling schema could also be applied to the candidate mitigation sites once site boundaries have been delineated.

Incorporate Monitoring of Effectiveness of Actions. The sampling schema for an implementation action follows the criterion from the previous paragraph, with the sample population based on the geospatial footprint of the project area and the addition of control sites to determine effectiveness of the action. Control sites are chosen outside of the action area based on similarity of soils and existing vegetation community in the action area. Control sites can be a selection from existing statistically valid monitoring efforts such as the long-term monitoring sites that are a part of the BLM Landscape Monitoring Framework.

To account for the variability among sites of similar potential, a minimum of three control sites are selected for each strata present in the treatment area. Sample sufficiency analysis will be conducted after the first year of sampling to examine indicator variability within each stratum to determine if additional sites are needed in the implementation action or control areas.

Integrate Remote Sensing Monitoring Technologies. Considerable work has been done to develop methodologies for processing and analyzing remote sensing data in order to extract information suitable for assessing changes in certain environmental conditions over time. The AIM Strategy emphasizes the value of integrating remote sensing technologies into long-term monitoring programs, wherever feasible, in order to provide cost-effective methods for collecting data and analyzing effects (Toevs et al. 2011).

Remote sensing technologies provide several benefits. They support the collection of spatially comprehensive datasets that are not otherwise readily available. In addition, the collection of data from a satellite or aircraft is nonintrusive, a very valuable feature for assessing ecologically and culturally sensitive areas. Semi-automated data processing of remotely sensed images can be a cost-effective way to reliably detect and identify features and quantify parameters over large areas more frequently. This feature is desirable for monitoring spatially heterogeneous and temporally dynamic arid and semiarid environments. Historic archives of remotely sensed data permit retrospective assessments and are thus suitable for long-term monitoring (Washington-Allen et al. 2006).

The limitations of remote sensing are that such measurements are indirect, and the spatial sampling unit (i.e., pixel) is arbitrary. In remote sensing, spectral reflectance signals from elements on the ground are assumed to be isolated from environmental and instrumental noise (Stow 1995). Further, targets are assumed to be spectrally separable from background, and different target types are assumed to have unique spectral signatures (Friedl, McGwire, and McIver 2001). The BLM interdisciplinary team should consult the AIM Strategy guidance and remote sensing experts to investigate cost-effective ways to incorporate the use of remote sensing technologies into the monitoring of mitigation actions.

Step 4. Develop Analysis and Reporting System

Interpreting the data to determine the status, departure, or rate of change requires comparison of data collected via field sampling and/or remote sensing against indicators of ecological attributes for reference conditions. These reference conditions will be based upon site or landscape potential which is described in ecological site descriptions or documented through reference sites. Ecological sites, or groupings of sites with similar ecological characteristics, are the basis for the monitoring schema because they react similarly to factors like disturbance or degradation (historic or current), which can lead to alternative stable plant communities outside the historic potential of the site. For this reason, ecological groupings are a basic unit for analysis and reporting. Elements of an ecological description that are helpful for defining reference conditions and interpreting departure from reference conditions include: state-and-transition conceptual models of plant community changes in response to disturbance or management; descriptions of the range of plant communities that could exist on the site in addition to the potential vegetation; descriptions of anthropogenic and natural disturbances and their potential to cause changes in plant communities; descriptions of dynamic soil properties (e.g., organic matter content, soil aggregate stability); and amount of bare ground. Report frequency will be established at the time the mitigation and monitoring actions are selected. Reports would be made publicly available through various media (e.g., available on public websites).

Step 5. Define Adaptive Management Approach

The BLM will use information derived from the Arizona monitoring plan to determine if resource management objectives described in the Lake Havasu, Yuma, and Lower Sonoran RMPs for the Arizona SEZs, the 2-mile buffer zone around each SEZ, and the areas where regional compensatory mitigation actions will occur are being met. If the objectives are not being met, the monitoring program information will be used to make necessary management adjustments to the mitigation actions. Reporting at multiple scales will inform decision makers on the effectiveness of management and mitigation actions, opportunities for adaptive management (e.g., adjusting operational parameters, modifying mitigation actions, and/or adding new mitigation actions), refinement of conceptual models, and evaluation of the monitoring program itself. Changes identified through adaptive management will be subject to environmental analysis, land use planning, and public involvement, as appropriate.

2.10 Implementation Strategy

This SRMS considered impacts that are likely to occur with the full build-out of each of the Arizona SEZs identified in the Solar PEIS and RDEP RODs. The AZ IDT found that while many potential impacts can be avoided and/or minimized, several residual impacts are likely to remain and may warrant compensatory mitigation as listed in Section 2.4.3.2.

Any authorized mitigation activities will be intended to provide mitigation through the duration of the project impacts with intensive monitoring and adaptive management for 50 years.[18] This extended time period is critical for effective implementation of mitigation. The proposed mitigation sites and actions will offset anticipated impacts of solar development in the Arizona SEZs and allow the BLM to sustain the yield of impacted resources for present and future generations.

The findings and recommendations offered here are intended to inform the decision-making process associated with leasing land in the Arizona SEZs for utility-scale solar energy development. At the discretion of the BLM authorized officer, all or part of these recommendations should be included in applicable NEPA analyses and the decision-making process.

While this SRMS focuses on solar energy development in the three SEZs in Arizona, the process outlined also can be applied to utility-scale renewable energy development in the REDAs that were designated in the RDEP ROD. In addition, background information regarding the ecoregion and nominated candidate sites are also available for use, if found to be relevant.

---

[18] Fifty years was assumed for the purposes of costing the mitigation obligation. The duration of project impacts may likely extend beyond the 30-year time frame of the project authorization.

3 REFERENCES

BLM (Bureau of Land Management), 2007, *Record of Decision and Lake Havasu Field Office Approved Resource Management Plan*. U.S. Department of the Interior, Bureau of Land Management, Lake Havasu Field Office, Lake Havasu City, AZ.

BLM 2008, National Environmental Policy Act. BLM Handbook H-1790-1. Bureau of Land Management, Washington, DC.

BLM 2010, *Yuma Field Office Record of Decision and Approved Resource Management Plan*. U.S. Department of the Interior, Bureau of Land Management, Yuma Field Office, Yuma, AZ.

BLM 2012a, *Approved Resource Management Plan Amendments/Record of Decision (ROD) for Solar Energy Development in Six Southwestern States*. U.S. Department of the Interior, Bureau of Land Management, Washington, DC.

BLM 2012b, *Lower Sonoran Record of Decision and Approved Resource Management Plan*. U.S. Department of the Interior, Bureau of Land Management, Lower Sonoran Field Office, Phoenix, AZ.

BLM 2012c, *Sonoran Desert Rapid Ecoregional Assessment Report*. U.S. Department of the Interior, Bureau of Land Management, Denver, CO.

BLM 2012d, *Final Environmental Impact Statement for the Restoration Design Energy Project*. U.S. Department of the Interior, Bureau of Land Management, Arizona State Office.

BLM 2013a, *Record of Decision and Approved Resource Management Plan Amendments for the Restoration Design Energy Project*. U.S. Department of the Interior, Bureau of Land Management, Arizona State Office.

BLM 2013b, *Interim Policy, Draft – "Regional Mitigation" Manual Section – 1794*. Instruction Memorandum No. 2013-142. U.S. Department of the Interior, Bureau of Land Management, Washington, DC.

BLM (Bureau of Land Management), 2014a, *Draft Procedural Guidance for Developing Solar Regional Mitigation Strategies*. U.S. Department of the Interior, Bureau of Land Management, Washington Office, Washington, DC. July. Available at: http://blmsolar.anl.gov/sez/policies/regional/.

BLM 2014b, *Solar Regional Mitigation Strategy for the Dry Lake Solar Energy Zone*. Technical Note 444. U.S. Department of the Interior, Bureau of Land Management, Southern Nevada District Office, Las Vegas, NV.

BLM and DOE (U.S. Department of Energy), 2012, *Final Programmatic Environmental Impact Statement (PEIS) for Solar Energy Development in Six Southwestern States*. FES 12-24, DOE/EIS-0403. U.S. Department of the Interior, Bureau of Land Management and U.S. Department of Energy.

DOI (Department of the Interior), 2015. Departmental Manual Part 600 (Public Land Policy), Chapter 6: Implementing Mitigation at the Landscape-Scale, Office of Policy Analysis. October 23. Available at: http://elips.doi.gov/elips/DocView.aspx?id=4209&searchid=1c0ac19b-efc7-42f9-ba57-e59278532005&dbid=0.

Elzinga, C.L., D.W. Salzer, and J.W. Willoughby, 1998, *Measuring and Monitoring Plant Populations.* Technical Reference 1730-1. U.S. Department of the Interior, Bureau of Land Management, National Business Center, Denver, CO. http://www.blm.gov/nstc/library/pdf/MeasAndMon.pdf.

Friedl, M.A., K.C. McGwire, and D.K. McIver, 2001, *"An Overview of Uncertainty in Optical Remotely Sensed Data for Ecological Applications,"* p. 258-283. In: C.T. Hunsaker, M.F. Goodchild, M.A. Friedl, and T.J. Case (eds.). *Spatial Uncertainty in Ecology: Implications for Remote Sensing and GIS Applications,* Springer, NY.

Lower Colorado River Multi-Species Conservation Program (LCRMSCP), 2004, *Lower Colorado River Multi-Species Conservation Program, Volume II: Habitat Conservation Plan.* Final. December 17. (J&S 00450.00.) Sacramento, CA.

MacKinnon, W.C., J.W. Karl, G.R. Toevs, J.J. Taylor, M. Karl, C.S. Spurrier, and J.E. Herrick, 2011, *BLM Core Terrestrial Indicators and Methods.* Technical Note 440. U.S. Department of the Interior, Bureau of Land Management, National Operations Center, Denver, CO.

Pellant, M., P. Shaver, D.A. Pyke, and J.E. Herrick. 2005. Interpreting Indicators of Rangeland Health, Version 4. Tech Ref 1734–6. Bureau of Land Management, National Science and Technology Center, Denver, CO. http://www.blm.gov/nstc/library/techref.htm.

Stow, D.A, 1995, *"Monitoring Ecosystem Response to Global Change: Multitemporal Remote Sensing Analysis,"* p. 254–286. In: J.M. Moreno et al. (eds.) Global Change and Mediterranean-Type Ecosystems. Springer-Verlag New York, Inc.

Toevs, G.R., J.J. Taylor, C.S. Spurrier, W.C. MacKinnon, and M.R. Bobo, 2011, *Assessment, Inventory, and Monitoring Strategy for Integrated Renewable Resources Management.* U.S. Department of the Interior, Bureau of Land Management, National Operations Center, Denver, CO.

USGS National Gap Analysis Program, 2004, Provisional Digital Land Cover Map for the Southwestern United States. Version 1.0. RS/GIS Laboratory, College of Natural Resources, Utah State University.

Washington-Allen, R.A., N.E. West, R.D. Ramsey, and R.A. Efroymson, 2006, *"A Protocol for Retrospective Remote Sensing-Based Ecological Monitoring of Rangelands."* Rangeland Ecology and Management 59 (1): 19–29.

4 GLOSSARY

Adaptive management: a system of management practices based on clearly identified outcomes and monitoring to determine whether management actions are meeting desired outcomes and; if not, facilitating management changes that will best ensure that outcomes are met or re-evaluated. Adaptive management recognizes that knowledge about natural resource systems is sometimes uncertain.

Additionality: improves the baseline conditions of the impacted resource, and is demonstrably new and would not have occurred without the compensatory mitigation action.

Avoidance: avoiding the impact altogether by not taking a certain action or parts of an action (40 CFR 1508.20(a)).

Baseline: the pre-existing condition of a resource, at all relevant scales, which can be quantified by an appropriate attribute(s). During environmental reviews, the baseline is considered the affected environment that exists absent the project's implementation, and is used to compare predictions of the effects of the proposed action or a reasonable range of alternatives.

Best management practices (BMPs): state-of-the-art, efficient, effective, and practicable mitigation measures for avoiding, minimizing, rectifying, and reducing or eliminating impacts over time. BMPs for solar development in Arizona are identified in BLM's Western Solar Plan and Restoration Design Energy Project.

Change agents: an environmental phenomena or human activity that can alter or influence the future condition and/or trend of a resource. Some change agents (e.g., roads) are the result of direct human actions or influence; others (e.g., climate change, wildland fire, and invasive species) may involve natural phenomena or be partially or indirectly related to human activities.

Coarse filter: elements such as vegetation communities, ecosystems, or land classes for planning and management across landscape- and regional-level management units.

Compensation: compensating for the impact by replacing or providing substitute resources or environments (40 CFR 1508.20(e)).

Compensatory mitigation action: an activity, process, or measure that may include restoration, establishment, enhancement, and preservation of resources offsetting residual effects.

Compensatory mitigation obligation: the compensatory mitigation actions required by the BLM to mitigate residual effects to resources from a land use activity, or fees paid to BLM or other entities to be used to mitigate residual effects to resources from a land use activity.

Compensatory mitigation site: the areas where compensatory mitigation actions are located.

Conservation elements: resources with regional conservation importance, including: species, species assemblages, ecological systems, habitats, physical resources (e.g., air, soils, and hydrology), cultural resources, and visual resources.

Design features: required measures or procedures incorporated into the proposed action or alternatives which could avoid, minimize, mitigate, or otherwise reduce adverse impacts of a project proposal. Design features for solar development in Arizona are identified in BLM's Western Solar Plan and Restoration Design Energy Project.

Durability: a state in which the measurable environmental benefits of mitigation will be sustained, at minimum, for as long as the associated harmful impacts of the authorized activity continue. The "durability" of a mitigation measure is influenced by: (1) the level of protection or type of designation provided; and (2) financial and long-term management commitments.

Duration of the impact: the temporal extent of resource impacts resulting from permitted actions. The duration of some impacts may be indefinite or perpetual.

Effective: produces the desired outcome.

Effects: the adverse direct, indirect, and cumulative impacts from a land use activity; effects and impacts as used in this document are synonymous.

Enhancement: the manipulation of resources to heighten, intensify, or improve a specific resource.

Fine filter: meant to complement the coarse filter by targeting species with requirements that will not be met through the broad brush of dominant vegetation communities—rare, threatened or endangered species, wildlife species of management interest, or those species that consistently use ecotones or multiple habitats on a diurnal or seasonal basis.

Goal (regional goal or land use plan goal): a broad statement of a desired outcome. Goals are usually not quantifiable and may not have established time frames for achievement.

Impacts: the adverse direct, indirect, and cumulative effects from a land use activity; effects and impacts as used in this document are synonymous.

Landscape: a geographic area encompassing an interacting mosaic of ecosystems and human systems that is characterized by a set of common management concerns. The landscape is not defined by the size of the area, but rather by the interacting elements that are relevant and meaningful in a management context.

Minimization: minimizing impacts by limiting the degree or magnitude of the action and its implementation (40 CFR 1508.20(b)).

Mitigation: includes, avoiding the impact altogether by not taking a certain action or parts of an action; minimizing impacts by limiting the degree or magnitude of the action and its implementation; rectifying the impact by repairing, rehabilitating, or restoring the affected environment; reducing or eliminating the impact over time by preservation and maintenance operations during the life of the action; and, compensating for the impact by replacing or providing substitute resources or environments (40 CFR 1508.20).

Mitigation Desired Outcome: a clearly-defined and measurable result of a compensatory mitigation action.

Mitigation hierarchy: see *Mitigation*, the process and order of preference for the application of mitigation, i.e., avoidance, minimization, remediation, reduction over time, and/or compensation, in order.

Mitigation Strategy: a document that identifies, evaluates, and communicates potential mitigation needs and mitigation measures in a geographic area, at relevant scales, in advance of anticipated land use activities.

NEPA process/analysis: analysis prepared pursuant to the National Environmental Policy Act, such as a planning- or project-level environmental assessment (EA) or environmental impact statement (EIS).

No net loss: when mitigation results in no negative change to baseline conditions (e.g., fully offset or balanced).

Objective (regional objective or land use plan objective): a description of a desired outcome for a resource in a land use plan. Objectives can be quantified and measured and, where possible, have established time frames for achievement.

On-site Mitigation: mitigation implemented in the project area.

Operations and Maintenance: a budgeting term including costs of operation and maintenance of, for example, a mitigation feature.

Performance Monitoring: Short-term monitoring of the restoration effort success. In this SRMS, it refers to a five-year initial implementation time period.

Preservation: the removal of a threat to, or preventing the decline of, resources. Preservation may include the application of new protective designations on previously unprotected land or the relinquishment or restraint of a lawful use that adversely impacts resources.

Proponent-responsible compensatory mitigation: resources that are restored, established, enhanced, and/or preserved, by an authorized land user (or an authorized agent or contractor), for the purpose of compensating for residual effects to resources from land use activities.

Residual impacts: any adverse reasonably foreseeable effects that remain after the application of the first four steps in the mitigation hierarchy; also referred to as unavoidable impacts.

Resources (and their values, services, and/or functions): *resources* are natural, social, or cultural objects or qualities; *resource values* are the importance, worth, or usefulness of resources; *resource services* are the benefits people derive from resources; and *resource functions* are the physical, chemical, and/or biological processes that involve resources.

Restoration: the manipulation of degraded resources in order to return the resources to an un-degraded condition.

This page intentionally left blank

APPENDIX A

Impact Assessment Summary Tables

This page intentionally left blank

The following tables summarize the Bureau of Land Management and Argonne National Laboratory subject matter expert responses to the process steps and criteria used to identify the unavoidable impacts that are likely to occur as a result of solar development in the Arizona Solar Energy Zones. The process steps and criteria for identifying unavoidable impacts are outlined in Section 2.4.3.1 of this document.

Table A-1. Agua Caliente Solar Energy Zone: Impact Assessment Summary Table

In Yuma County in southwestern Arizona, Yuma Field Office – 1,617 developable acres, up to 328 MW generation capacity, assuming 80% development. Sources: Draft and Final Restoration Design Energy Project (RDEP) EIS containing analysis for Agua Caliente SEZ (available at: http://www.blm.gov/az/st/en/prog/energy/arra_solar/DEIS.html) and Draft and Final Solar PEIS (available at: http://blmsolar.anl.gov/)

Resource/Issue	Agua Caliente Solar Energy Zone Impacts[1]	On-site Mitigation[2] Avoidance	On-site Mitigation[2] Minimization	Residual Adverse Impacts?[3]
Acoustics Section 4.2.12[4]	**Direct:** Increased noise levels during construction, operations, and decommissioning may be experienced but would not exceed regulatory levels. **Indirect:** Construction noise from the SEZ is not likely to adversely affect any of the nearby specially designated areas. **Cumulative:** Nearest residents live in Hyder, approximately 7 mi (11 km) from the SEZ, thus cumulative noise effects during the construction or operation of solar facilities are unlikely. **Data Gaps[5]:** Refined modeling would be warranted along with background noise measurements during project-specific assessments.	Solar facilities must be located far enough away from residences, or include engineering and/or operational methods such that county, state, and/or federal regulations for noise are not exceeded. See programmatic design features at http://blmsolar.anl.gov/documents/docs/peis/programmatic-design-features/Noise.pdf and SEZ-specific design features in the RDEP ROD, Table B-4.	Increases in noise will be limited to less than a 10 dBA increase above ambient levels, and will not exceed local noise standards. The hours of daily activities will be limited and noise barriers will be constructed if needed and practicable. Coordination with nearby residents is recommended. See programmatic and SEZ-specific design features.	Maybe (depends on technology and engineering controls). Generally impacts from solar development are expected to be temporary, localized, and readily mitigated onsite.

[1] The impacts assessment assumed 80% of the developable SEZ area will be used for solar development.
[2] These columns give examples of avoidance and minimization measures that are specified in the Record of Decision for the Final Solar PEIS and will be required. Additional avoidance and minimization measures proposed by the BLM Interdisciplinary Team are listed and should be evaluated through project-specific environmental analyses. Monitoring is planned to verify the implementation and effectiveness of avoidance and minimization measures.
[3] Residual or unavoidable impacts are residual effects that cannot be adequately mitigated onsite by avoidance and/or minimization.
[4] All section numbers in this Agua Caliente table are from the Final RDEP EIS, unless otherwise indicated.
[5] Data gaps have not been identified for all resources in this table. Additional data gaps may be identified during future SEZ- or project-specific assessments.

Regional Mitigation Strategy for the Arizona SEZs

Table A-1. (Cont.)

Resource/Issue	Agua Caliente Solar Energy Zone Impacts[1]	On-site Mitigation[2] — Avoidance	On-site Mitigation[2] — Minimization	Residual Adverse Impacts?[3]
Air Quality Section 4.2.1	**Direct:** Fugitive dust and equipment exhaust emissions during construction could result in exceedance of Ambient Air Quality Standards (AAQS) for particulate matter (PM) at SEZ boundaries. Specifically, 24-hour PM_{10} and 24-hour and annual $PM_{2.5}$ concentrations could exceed AAQS at the SEZ boundaries and in the immediate surrounding areas during construction of solar facilities. High PM_{10} concentrations would be limited, however, to the immediate areas surrounding the SEZ boundary and would decrease quickly with distance. Generation of fugitive dust may result in exposure to respirable particulates and/or microbes (human health impacts). The majority of the soils on the SEZ have been characterized as having high potential for wind erosion. **Indirect:** Decreased visibility in nearby residential or specially-designated areas due to elevated PM levels from soil disturbance/grading during construction. **Cumulative:** Cumulative effects due to dust emissions would be greatest if multiple solar projects had overlapping construction periods. **Data Gaps:** Predicted 24-hour PM_{10} and 24-hour and annual $PM_{2.5}$ concentration levels not included in RDEP analysis. Monitoring for PM during all phases of development will be required to identify levels exceeding AAQS.	See programmatic design features at http://blmsolar.anl.gov/documents/docs/peis/programmatic-design-features/Air_Quality_Climate.pdf and SEZ-specific design features in the RDEP ROD, Table B-4.	All soil disturbance activities and travel on unpaved roads will be suspended during periods of high winds. A critical site-specific wind speed will be established based on soil properties determined during site characterization, and wind speed monitoring will be required at the site during construction, operation, and reclamation. Dust suppression measures will be implemented during all phases of development (construction, operations, and decommissioning). See programmatic and SEZ-specific design features. Also recommend evaluation of solar panel mounting and other disturbance minimizing technologies in project-level NEPA alternatives (e.g., no grading of the site, retention of maximum native vegetation, use of low emission vehicles, placing gravel on roads, use of "drive and crush" installation). Recommend re-vegetation of the SEZ with native vegetation to increase soil stability as a plan of development feature to further minimize the amount of grading and surface disturbance and promote reduced dust emissions and PM levels.	Maybe (if site is graded). Level of site grading and disturbance to native vegetation would be primary driver of residual impact for full build-out of SEZ. Impacts are not expected to result in noncompliance with National Air Quality Standards.

Table A-1. (Cont.)

Resource/Issue	Agua Caliente Solar Energy Zone Impacts[1]	On-site Mitigation[2]		Residual Adverse Impacts?[3]
		Avoidance	Minimization	
Climate Change Section 5.11.4 of Draft Solar PEIS for soil storage capacity	**Direct:** Possible impact through loss of carbon storage capacity of the soil (estimated at 100 g carbon/m^2). Preliminary calculations show loss of CO_2 storage capacity as 1.6 tons/acre/yr (3,264 tons/yr for SEZ full build-out). **Positive impact:** Solar power generation reduces demand for energy from fossil fuels, and thereby reduces greenhouse gas emissions (emissions avoided not given in RDEP EIS, would be similar to emissions avoided for the similarly-sized Gillespie SEZ (i.e., from about 347,000-624,000 tons/yr CO_2 avoided at full build-out depending on technology). **Cumulative:** Over the long-term, the development of solar energy may contribute to reduced greenhouse gas emissions (if the development offsets electricity generation by fossil fuel plants). About 65% of electricity in AZ is produced in fossil fuel plants. Based on data from the Sonoran Desert Rapid Ecoregional Assessment (REA), the SEZ is situated in an area with moderate potential for future climate change (e.g., increased temperature, decreased precipitation, and changes in vegetation and habitat).	Native vegetation cover and soils will be maintained and grading will be minimized. See programmatic design features for vegetation at http://blmsolar.anl.gov/documents/docs/peis/programmatic-design-features/Ecological_Resources.pdf and SEZ-specific design features in the RDEP ROD, Table B-4.	See programmatic and SEZ-specific design features.	No. Impacts are likely to be positive. No mitigation likely needed.
Cultural Section 4.2.3	**Direct:** Development may adversely affect cultural resources. There is potential to physically impact prehistoric and historic sites and features. There could be impacts on views from the Juan Bautista de Anza National Historic Trail and the Sears Point Area of Critical Environmental Concern. **Indirect:** Erosion impacts on the cultural landscape outside of the SEZ resulting from land disturbances and modified hydrologic patterns; increased accessibility and potential for damage to eligible sites in the non-development area as well as outside of the SEZ. **Cumulative:** Eligible sites and cultural landscapes are present and could be impacted in the SEZ and adjacent areas. There are large World War II military training ranges in and near the SEZ that have the potential to be affected. **Data Gaps:** Documentation of a 100% pedestrian archaeological survey of the SEZ is currently being completed. The Section 106 consultation process must also be completed at the project level and has the potential to result in additional information to consider.	Significant resources clustered in specific areas which retain sufficient integrity will be avoided to the extent possible. See programmatic design features at http://blmsolar.anl.gov/documents/docs/peis/programmatic-design-features/Cultural.pdf and SEZ-specific design features in the RDEP ROD, Table B-4.	A recently completed archaeological survey has informed the creation of non-development areas within SEZ. An agreement document and a Historic Property Treatment Plan will be written pursuant to Section 106 for the resolution of adverse effects to any historic property included in or eligible for inclusion in the National Register of Historic Places. See programmatic and SEZ-specific design features.	Yes. Impacts on non-renewable resources are both irretrievable and irreversible. Tribal consultation may present situations where data recovery or collection onsite is not possible. Procedures to handle inadvertent discoveries will be addressed in a monitoring and discovery plan developed during the permitting process.

Table A-1. (Cont.)

Resource/Issue	Agua Caliente Solar Energy Zone Impacts[1]	On-site Mitigation[2]		Residual Adverse Impacts?[3]
		Avoidance	**Minimization**	
	Direct: Development will adversely affect characteristic vegetation (e.g., creosote bush and white bursage) through destruction and loss of habitat. Development, including vegetation removal, land clearing, grading, changes in surface water flow, and dust deposition may alter soils and vegetation communities and result in the establishment of invasive species and noxious weeds within the SEZ.	Dry wash, dry wash woodland, saguaro cactus, and ironwood (including those outside of washes) vegetation communities within the SEZ will be avoided to the extent practicable. A buffer area will be maintained around dry washes and dry wash woodland habitats to reduce the impact potential.	Appropriate engineering controls will be used to minimize impacts resulting from surface water runoff, erosion, sedimentation, altered hydrology, accidental spills, or fugitive dust deposition to these habitats. Appropriate buffers and engineering controls will be determined through agency consultation.	Yes. Development would result in direct removal or disturbance of these native plant communities, special soil environments, and the ecosystem services they provide.
Ecology: Vegetation Section 4.2.21	**Indirect:** Loss of native vegetation due to dust deposition from construction, operations, and decommissioning, increased surface water runoff and related erosion, or through the introduction of invasive species. Establishment of noxious weeds in the SEZ may result in spread of weeds to adjacent areas	Travel through weed-infested areas will be avoided; vehicles and equipment will be inspected and cleaned to avoid spread of weeds; ground disturbance will be limited, creation of soil conditions that promote weed germination and establishment will be avoided, seed and plant parts will be disposed of.	Yucca species and most agave and cactus species will be salvaged prior to land clearing and transplanted in accordance with Arizona Native Plant Law.	
	Cumulative: Solar energy development could be a contributor to cumulative impacts on some vegetation communities, depending on the type, number, and location of other developments in the region.	See programmatic design features at http://blmsolar.anl.gov/documents/docs/peis/programmatic-design-features/Ecological_Resources.pdf and SEZ-specific design features in the RDEP ROD, Table B-4.	See programmatic and SEZ-specific design features.	

Table A-1. (Cont.)

Resource/Issue	Agua Caliente Solar Energy Zone Impacts[1]	On-site Mitigation[2]		Residual Adverse Impacts?[3]
		Avoidance	**Minimization**	
	Direct: Development will adversely affect characteristic vegetation (e.g., creosote bush, white bursage, cactus, paloverde, and ironwood) through destruction and loss of habitat. Development, including vegetation removal, land clearing, grading, changes in surface water flow, and dust deposition may alter soils and vegetation communities and result in the establishment of invasive species and noxious weeds within the SEZ.	Dry washes, playas, and wetlands within the SEZ and dry washes within the access road corridor will be avoided to the extent practicable. A buffer area will be maintained around wetlands, playas, and dry washes to reduce the potential for impacts.	See programmatic design features.	Maybe. Depends on the degree of avoidance and engineering controls. Development may alter ephemeral stream channels that can impact flooding and debris flows during storms, groundwater recharge, ecological habitats, and riparian vegetation communities.
Ecology: Riparian Areas Section 4.2.21	**Indirect:** Loss of native vegetation due to dust deposition from construction and operations, increased surface water runoff and related erosion, or through the introduction of invasive species. Establishment of noxious weeds in the SEZ may result in their spreading to adjacent areas. **Cumulative:** Solar energy development could be a contributor to cumulative impacts on some vegetation communities, depending on the number and location of other developments in the region.	Appropriate engineering controls will be used to minimize impacts on dry wash, dry wash woodland and chenopod scrub, including downstream occurrence, resulting from surface water runoff, erosion, sedimentation, altered hydrology, accidental spills, or fugitive dust deposition to these habitats. Appropriate buffers and engineering controls will be determined through agency consultation. See programmatic design features at http://blmsolar.anl.gov/documents/docs/peis/programmatic-design-features/Ecological_Resources.pdf and SEZ-specific design features in the RDEP ROD, Table B-4.	Groundwater withdrawals will be limited to reduce the potential for dependent communities, such as, microphyll (paloverde/ironwood) communities, or riparian habitats along the Gila River.	Reductions to the connectivity of these areas with existing surface waters and groundwater could limit water availability and thus alter the ability of the area to support vegetation and aquatic species. This could reduce overall stability of the natural landscape.

Regional Mitigation Strategy for the Arizona SEZs

Table A-1. (Cont.)

Resource/Issue	Agua Caliente Solar Energy Zone Impacts[1]	On-site Mitigation[2] Avoidance	Minimization	Residual Adverse Impacts?[3]
Ecology: Invasive & Noxious Weeds Section 4.2.21	**Direct:** Development, including vegetation removal, land clearing, grading, changes in surface water flow, and dust deposition may alter soils and vegetation communities and result in the establishment of invasive species and noxious weeds within the SEZ. **Indirect:** Loss of native vegetation due to dust deposition from construction and operations, increased surface water runoff and related erosion, or through the introduction of invasive species. Establishment of noxious weeds in the SEZ may result in their spreading to adjacent areas. **Cumulative:** Solar energy development could be a contributor to cumulative impacts on some vegetation communities, depending on the number and location of other developments in the region.	Travel through weed infested areas will be avoided; vehicles and equipment will be inspected and cleaned to avoid spread of weeds; ground disturbance will be limited, creation of soil conditions that promote weed germination and establishment will be avoided, and disposal of seed and plant parts will be disposed of to reduce invasive species establishment. See programmatic design features at http://blmsolar.anl.gov/documents/docs/peis/programmatic-design-features/Ecological_Resources.pdf and SEZ-specific design features in the RDEP ROD, Table B-4.	Impacts will be minimized through development of a Weed Management Plan and use of weed-free seed to support re-vegetation efforts, control invasive species, and prevent increase in fires. See programmatic design features.	Maybe. On-site mitigation will reduce, but not eliminate, the potential for invasive species. The degree of disturbance creates a significant opportunity for the establishment of invasive species and weeds.
Ecology: Terrestrial Wildlife and Aquatic Biota Section 4.2.6	**Direct:** Loss of habitat and connectivity for several species of amphibians, reptiles, mammals, bats, and invertebrates. Ground disturbance, fugitive dust generated by project activities, lighting, vegetation clearing, spread of invasive species, accidental spills, harassment, and impacts on ephemeral washes could impact terrestrial wildlife within the SEZ. Impacts from noise on wildlife could occur, especially on bat species, if the SEZ is located near any bat roosts. **Indirect:** Outside the SEZ, impacts could occur from habitat loss or modification, increased human presence in the area, surface runoff, dust, noise, lighting, or accidental spills. **Cumulative:** Cumulative effects on some species could occur depending on the type, number, and location of other developments in the region. **Data Gaps:** Impacts on terrestrial wildlife from construction noise would have to be considered on a project-specific basis, especially for bat species.	Development will avoid wetlands, washes, and riparian areas identified during site-specific surveys. See programmatic design features at http://blmsolar.anl.gov/documents/docs/peis/programmatic-design-features/Ecological_Resources.pdf and SEZ-specific design features in the RDEP ROD, Table B-4.	The fencing around the solar energy development should not block the migratory corridors of mammals, particularly big game species. Big game habitat will be managed in coordination with Arizona Game and Fish Department management objectives and BLM Land Use Plan objectives. See programmatic and SEZ-specific design features.	Yes. Development of the Agua Caliente SEZ will likely impact up to 2,550 acres of wildlife habitat. Level of site grading and disturbance to native vegetation would be primary driver of residual impact for full build-out of SEZ.

Table A-1. (Cont.)

| Resource/Issue | Agua Caliente Solar Energy Zone Impacts[1] | On-site Mitigation[2] | | Residual Adverse Impacts?[3] |
		Avoidance	Minimization	
Ecology: Migratory Birds Section 4.2.6	**Direct:** Loss of individuals, habitat, and connectivity for several species protected under the Migratory Bird Treaty Act. Noise, lighting, and vegetation clearing could impact migratory birds using the SEZ. There is potential for birds to be attracted to solar fields (because they look like water) and impact with solar panels. Burning of wings in the solar radiation field between heliostats and power towers has been observed. There may also be impacts to night sky that may alter bird migratory behavior and habitat usage. Priority migratory bird species that may occur on or near the SEZ include Albert's towhee, Arizona Bell's vireo, Gila woodpecker, gilded flicker, LeConte's thrasher, Lincoln's sparrow, and Sprague's pipit.[19] **Indirect:** Outside the SEZ, impacts could occur from habitat loss or modification related to on-site disturbances (noise, lighting, habitat fragmentation). **Cumulative:** Impacts to migratory birds could occur; depending on the type, number, and location of other developments in the region. **Data Gaps:** Additional research needed on solar development impacts on migratory birds. Impacts on migratory birds from construction noise would have to be considered on a project-specific basis.	Effects to individual migratory birds and bird nests can be avoided by not constructing during the breeding season. Timing limitation should be enforced from May 15–July 15 for any surface disturbing activities to protect migratory bird nesting and brood rearing. If construction takes place during the breeding season, nest surveys will be conducted. See programmatic design features at http://blmsolar.anl.gov/documents/docs/peis/programmatic-design-features/Ecological_Resources.pdf and SEZ-specific design features in the RDEP ROD, Table B-4.	Recommend implementation of technologies that minimize the amount of reflective surfaces, or alter how the surfaces are perceived by wildlife, to reduce the "lake effect" in attracting migratory birds and other wildlife. See programmatic and SEZ-specific design features.	Yes. Development of the Agua Caliente SEZ will likely impact up to 2,550 acres of migratory bird habitat. Some level of bird injury/fatality has been observed for all types of solar facilities (through collisions with equipment or from burns). Research is ongoing to quantify impacts and identify effective mitigation measures.

[19] Priority migratory bird species for the SEZ were determined based on those species discussed in the Yuma RMP and the distribution of Arizona Natural Heritage Program tracked species and USFWS Birds of Conservation Concern in the Arizona Habimap tool (http://www.habimap.org/).

Table A-1. (Cont.)

| Resource/Issue | Agua Caliente Solar Energy Zone Impacts[1] | On-site Mitigation[2] | | Residual Adverse Impacts?[3] |
		Avoidance	Minimization	
Ecology: Plant Special Status Species Section 4.2.19	**Direct:** No Endangered Species Act (ESA)-listed plant species have been identified that have suitable habitat within the SEZ. Ground disturbance, land clearing and grading, fugitive dust generated by project activities, and the spread of invasive species may result in loss of special status plant species habitat, if present, and might result in loss of individual plants. Subsequent BLM analyses show that Agua Caliente SEZ is actually outside the occurrence area for Schott's wire lettuce. Schott's wire lettuce is a dune species and dune habitat is not found on the Agua Caliente SEZ. **Indirect:** Indirect impacts to individuals and habitat outside of the SEZ could occur from surface runoff, dust, or accidental spills. **Cumulative:** There would be no cumulative impacts on special status plant species unless they are discovered during pre-disturbance surveys (cumulative impacts then might be due to habitat destruction and overall development and fragmentation of the area). **Data Gaps:** Pre-disturbance surveys are required to identify the presence and abundance of special status species.	Based on data from pre-disturbance surveys, disturbance to occupied habitats would be avoided to the extent practicable. See programmatic design features at http://blmsolar.anl.gov/documents/docs/peis/programmatic-design-features/Ecological_Resources.pdf and SEZ-specific design features in the RDEP ROD, Table B-4.	See programmatic and SEZ-specific design features.	No, unless special status plant species are discovered during pre-disturbance surveys. There are no known SSS plant species within the SEZ.

Table A-1. (Cont.)

Resource/Issue	Agua Caliente Solar Energy Zone Impacts[1]	On-site Mitigation[2] Avoidance	On-site Mitigation[2] Minimization	Residual Adverse Impacts?[3]
	Direct: Ground disturbance, land clearing and grading, and fugitive dust generated by project activities would result in loss of special status animal species habitat, if present, and might result in loss of individual animals. Impacts from noise on special status wildlife could also occur. There is one ESA-listed species that may occur on or near the SEZ, the lesser long-nosed bat.[20] However, the USFWS has determined that effects to lesser long-nosed bat potential foraging habitat as a result of SEZ development would be extremely unlikely and discountable due to the distance of the SEZ from any known roost locations (>50 miles). Seven BLM sensitive species may occur on or near the SEZ (Cactus ferruginous pygmy-owl, golden eagle, Le Conte's Thrasher, western burrowing owl, California leaf-nosed bat, Pale Townsend's big-eared bat, and Yuman desert fringe-toed lizard). Subsequent BLM analyses show that the Agua Caliente SEZ is actually outside the occurrence area of the following species: Cactus ferruginous pygmy-owl, golden eagle, Yuman desert fringe-toed lizard.		If avoidance is not possible for some species, translocation of individuals from areas of direct effects or compensatory mitigation may be employed.	
Ecology: Animal Special Status Species Section 4.2.19	No Category 1, 2, or 3 desert tortoise habitat has been identified within the SEZ; however, Category 3 desert tortoise habitat occurs outside the SEZ to the north and northwest. Desert tortoises may still occur in lower quality habitat on the SEZ where they may be directly impacted by solar development.

The SEZ is within the non-essential experimental population area for the Sonoran pronghorn. | Compliance with the Bald & Golden Eagle Protection Act would be ensured and Eagle Take Guidance would be followed (if necessary).

Based on data from pre-disturbance surveys, disturbance to suitable habitats will be avoided to the extent practicable.

See programmatic design features at http://blmsolar.anl.gov/documents/docs/peis/programmatic-design-features/Ecological_Resources.pdf and SEZ-specific design features in the RDEP ROD, Table B–4. | Regarding avoidance and minimization onsite, consultation with the USFWS will be conducted to address the potential for impacts on ESA-listed and proposed species and to identify mitigation measures for implementation.

Groundwater withdrawals will be avoided or minimized to reduce or eliminate impacts on nine special status species.

See programmatic and SEZ-specific design features. | Yes. Animal SSS along with other wildlife represent a basic component of the ecosystem.

Level of site grading and disturbance to native vegetation would be the primary driver of residual impact to functional habitat for full build out of the SEZ. |
	Indirect: Indirect impacts could occur from habitat loss or modification related to habitat fragmentation, surface runoff, dust, noise, lighting, or accidental spills.			
	Cumulative: There could be cumulative impacts on some special status animal species due to habitat destruction and overall development and fragmentation of the area.			
	Data Gaps: Pre-disturbance surveys are required to identify the presence and abundance of special status species.			

[20] Based on ESA Section 7 consultation with the USFWS. Conservation recommendations were provided by the USFWS for the non-essential experimental populations of Sonoran pronghorn.

Regional Mitigation Strategy for the Arizona SEZs

Table A-1. (Cont.)

Resource/Issue	Agua Caliente Solar Energy Zone Impacts[1]	On-site Mitigation[2]		Residual Adverse Impacts?[3]
		Avoidance	Minimization	
Environmental Justice Section 4.2.5	**Direct:** There are minority populations within 25 miles (40 km) of the SEZ, so any adverse impacts of solar projects could affect these populations. No low-income populations were identified within this area. Farm workers live near Hyder seasonally. **Indirect:** None identified. **Cumulative:** Contributions from solar development in the SEZ would likely be small and would not be expected to significantly contribute to cumulative impacts on minority populations.	See programmatic design features at http://blmsolar.anl.gov/documents/docs/peis/programmatic-design-features/Environmental_Justice.pdf and SEZ-specific design features in the RDEP ROD, Table B-4.	If possible, public relations materials should be available in Spanish due to the large Hispanic population in the area. See programmatic and SEZ-specific design features.	Maybe
Hydrology: Surface Water Section 4.2.23	**Direct:** Land clearing, land leveling, vegetation removal, and spills and runoff associated with development of the SEZ could increase surface runoff, reduce infiltration/recharge, cause loss of ephemeral stream networks, cause a reduction in evapotranspiration rates, increase sediment transport (by water), change sediment transport (by wind), and degrade water quality. Based on an evaluation of data in the Sonoran Desert REA, ephemeral drainages with high potential for water erosion occur on the SEZ. **Indirect:** Indirect impacts from development on ephemeral and perennial surface water features could occur. **Cumulative:** Alterations to ephemeral stream networks can alter groundwater recharge and surface runoff processes potentially impacting the basin-scale water balance and water quality aspects of water features receiving surface runoff. **Data Gaps:** Project siting and design will need to consider impacts to the stream channels and washes located in the SEZ.	Any projects impacting a wash or stream channel that are classified as a jurisdictional water of the United States will require coordination/permitting through the US Army Corps of Engineers. See programmatic design features at http://blmsolar.anl.gov/documents/docs/peis/programmatic-design-features/Water.pdf and SEZ-specific design features in the RDEP ROD, Table B-4.	Required measures should minimize sheet flow. See programmatic and SEZ-specific design features.	Yes. Hydrology is a basic component of the ecosystem. Reconfiguration of topography for solar development would have residual impacts to surface hydrology with potential impacts on other resources, including vegetation.

Regional Mitigation Strategy for the Arizona SEZs

Table A-1. (Cont.)

Resource/Issue	Agua Caliente Solar Energy Zone Impacts[1]	On-site Mitigation[2]		Residual Adverse Impacts?[3]
		Avoidance	Minimization	
Hydrology: Water Quality and Groundwater Availability Section 4.2.23	**Direct:** Groundwater withdrawals for development may cause declines in groundwater elevations that can impact water availability for surface water features, vegetation, ecological habitats, regional groundwater flow paths, and other groundwater users in the basin. The SEZ is located in the Lower Gila Basin. **Indirect:** Groundwater withdrawals for solar energy facilities have the potential to affect other groundwater users in the basin. **Cumulative:** Cumulative impacts on groundwater could occur when combined with other future developments in the region.	Groundwater analyses suggest that full build-out of wet-cooled technologies is not feasible. See programmatic design features at http://blmsolar.anl.gov/documents/docs/peis/programmatic-design-features/Water.pdf and SEZ-specific design features in the RDEP ROD, Table B-4.	Industrial water use limited to solar photovoltaic, solar thermal with dry-cooling, or similar low-water use technologies. See programmatic and SEZ-specific design features.	Maybe. It is possible for impacts on groundwater aquifers to be avoided or minimized.
Lands & Realty Section 4.2.8	**Direct:** Development of the SEZ could disturb 2,550 acres (10.3 km²). Development may require additional transmission and/or substation capacity. The SEZ is adjacent to a 290-MW PV solar facility in operation on private land. The large-capacity Hassayampa to North Gila transmission line passes within 0.5 mile of the southern end of the SEZ, and a new parallel 500-kv transmission line is expected to be in service by late 2014. **Indirect:** None identified. **Cumulative:** Projects within the SEZ would make only a small contribution to cumulative impacts because of its relatively small size.	See programmatic design features at http://blmsolar.anl.gov/documents/docs/peis/programmatic-design-features/Lands_and_Realty.pdf and SEZ-specific design features in the RDEP ROD, Table B-4.	Any potential impacts on the existing county road should be discussed with the county. See programmatic and SEZ-specific design features.	No. By regulation, any new activity must occur in deference to existing rights. Thus, potential impacts have been avoided.
Livestock Grazing Section 4.2.9	The SEZ is within the former Palomas Grazing Allotment. That allotment was made unavailable to livestock grazing in the January 2010 Yuma Field Office Resource Management Plan revision, as was the White Wing Allotment adjacent to the SEZ. There are no expected impacts to livestock grazing from solar development within the SEZ.	Not applicable	Not applicable	No

Regional Mitigation Strategy for the Arizona SEZs

Table A-1. (Cont.)

Resource/Issue	Agua Caliente Solar Energy Zone Impacts[1]	On-site Mitigation[2]		Residual Adverse Impacts?[3]
		Avoidance	Minimization	
Military & Civilian Aviation	**Direct:** The SEZ is within the visual corridor of a military training route (MTR) with a 300-foot (91-m) above-ground-level operating limit. Additionally, the Barry M. Goldwater Air Force Range is approximately 13 miles (21 km) south of the SEZ. The U.S. Army Yuma Proving Ground is approximately 7.5 miles (12 km) west of the SEZ. The development of any solar energy or transmission facilities that encroach into military airspace could interfere with military training activities. **Indirect:** None identified. **Cumulative:** Solar development occurring throughout the region, which is largely undeveloped, could result in small cumulative effects on the system of military training routes. Such effects would be limited by mitigations developed in consultation with the military.	See programmatic design features at http://blmsolar.anl.gov/documents/docs/peis/programmatic-design-features/Military_Civilian_Aviation.pdf	Coordination with Federal Aviation Administration and the military will be required on a project–specific basis to ensure that solar facilities do not interfere with operations. See programmatic design features.	Maybe (with respect to MTRs). Residual impacts will be evaluated based on coordination with the military and project-level NEPA.
Minerals Section 4.2.4	**Direct:** There are no documented proven oil and gas reserves in the SEZ. No high or moderate temperature geothermal resources exist in the SEZ, and there are no geothermal leases. There are no active mining claims within the SEZ, nor are there any active mines. The SEZ is in an area open for the disposal of salable minerals and is designated as having moderate potential for salable minerals, including sand, gravel, aggregate, cinders, decorative rock, and building stones. The BLM plans to withdraw the SEZ from mineral entry for a period of 20 years, precluding impacts from many types of mining activities, pending completion of a supporting environmental assessment. **Indirect:** None identified. **Cumulative:** None identified.	See programmatic design features at http://blmsolar.anl.gov/documents/docs/peis/programmatic-design-features/Mineral_Resources.pdf and SEZ-specific design features in the RDEP ROD, Table B-4.	See programmatic and SEZ-specific design features.	No

A-14

Regional Mitigation Strategy for the Arizona SEZs

Table A-1. (Cont.)

Resource/Issue	Agua Caliente Solar Energy Zone Impacts[1]	On-site Mitigation[2] Avoidance	Minimization	Residual Adverse Impacts?[3]
Native American Concerns Section 4.2.11	**Direct:** Concerns include noise, air quality, and visual resources. The SEZ is less than 10 mi from the Sears Point Area of Critical Environmental Concern (ACEC), a significant Native American heritage site. There may be visual, aural, or atmospheric intrusions. Traditional resource gathering areas may be impacted. Removal of cultural resources is a concern to tribes. **Indirect:** General habitat loss with vegetation clearing and water reduction that could affect species and ecosystem health. **Cumulative:** Development of solar energy facilities in combination with the development of other planned and foreseeable projects in the area would likely reduce the traditionally important plant and animal resources available to the tribes. Although some of these plant species are abundant, any level of impact may be of concern for the tribes. **Data Gaps:** Documentation of an archaeological survey of the entire SEZ is currently being completed and results will be shared with the tribes. Government-to-Government consultation for projects will be required to determine project-related issues of Native American concern.	Known human burial sites and rock art (panels of petroglyphs and/or pictographs) will be avoided. The BLM will consult with Indian tribes regarding the potential for unanticipated human remains and associated cultural items (as defined under the Native American Graves Protection and Repatriation Act) before a solar project is authorized. The purpose will be to discuss general guidance on treatment of cultural items. Springs and other water sources that are or may be sacred or culturally important, culturally important plant and wildlife species, and visual intrusion on sacred sites will be avoided to the extent practicable. EO13007 requires executive branch agencies to accommodate access to and ceremonial use of Indian sacred sites and to avoid adverse impacts to the physical integrity of such a site. Because solar facilities will be fenced and security procedures will limit or eliminate access, if a sacred site was declared, it may not be possible to mitigate impacts, other than through avoidance. See programmatic design features http://blmsolar.anl.gov/documents/docs/peis/programmatic-design-features/Native_American_Concerns.pdf and SEZ-specific design features in the RDEP ROD, Table B-4.	See programmatic and SEZ-specific design features. Mitigate onsite to comply with EO13007. BLM could facilitate the harvest of creosote prior to ground disturbance for instance, if identified as a concern.	Yes. Consultation on project applications will determine whether regional mitigation for Native American concerns is warranted.

Regional Mitigation Strategy for the Arizona SEZs

Table A-1. (Cont.)

Resource/Issue	Agua Caliente Solar Energy Zone Impacts[1]	On-site Mitigation[2] Avoidance	On-site Mitigation[2] Minimization	Residual Adverse Impacts?[3]
Paleontology Section 4.2.13	**Direct:** The SEZ includes 10 acres with geological units assigned to Potential Fossil Yield Classification Class 3; there are no Class 4 or 5 units within the SEZ. **Indirect:** None identified. **Cumulative:** Cumulative impacts would be dependent on whether significant resources are found within the SEZ and in additional project areas in the region. **Data Gaps:** A more detailed look at the geological deposits of the SEZ is needed to determine whether a paleontological survey is warranted for a specific project.	See programmatic design features at http://blmsolar.anl.gov/documents/docs/peis/programmatic-design-features/Paleo.pdf and SEZ-specific design features in the RDEP ROD, Table B-4.	The BLM will be notified immediately upon discovery of fossils. Work will be halted at the fossil site and continued elsewhere until qualified personnel, such as a paleontologist, can visit the site. He/she will determine if the site is significant and make recommendations for collection or other resource protection, if warranted. The use of training/education programs to reduce the amount of inadvertent destruction on paleontological sites could reduce the occurrence of human-related disturbance to nearby sites. See programmatic and SEZ-specific design features.	No. Design features will reduce the risk that any paleontological resources that are discovered will be destroyed.
Public Access and Recreation Sections 4.2.15 & 4.2.20	**Direct:** The SEZ is used for Off Highway Vehicle use, camping, and hunting. Development may preclude current recreational activities that occur within the SEZ boundary and potentially require rerouting of access to the Yuma East Special Recreation Management Area located to the north. **Indirect:** Indirect effects on recreation use would occur primarily on lands near the solar facilities and would result from the change in the overall character of undeveloped BLM-administered lands to an industrialized, developed area. People who are seeking more rural or primitive surroundings for recreation may have reduced or degraded recreational experiences. **Cumulative:** Multiple developments in the vicinity of the SEZ could cumulatively reduce recreational opportunities.	See programmatic design features at http://blmsolar.anl.gov/documents/docs/peis/programmatic-design-features/Public_Access_and_Recreation.pdf and SEZ-specific design features in the RDEP ROD, Table B-4.	Access to the Palomas Harquahala Road must be maintained or rerouted to maintain access to the Yuma East SRMA. Replacement of access lost for recreational use will be considered in project-specific. NEPA. See programmatic and SEZ-specific design features.	Yes

A-16

Regional Mitigation Strategy for the Arizona SEZs

Table A-1. (Cont.)

Resource/Issue	Agua Caliente Solar Energy Zone Impacts[1]	On-site Mitigation[2] Avoidance	On-site Mitigation[2] Minimization	Residual Adverse Impacts?[3]
Socioeconomics Section 4.2.16	**Direct:** Impacts to local economy as a result of expenditures of wages and salaries and the collection of state sales and income taxes. Development in the SEZ would create temporary construction jobs and permanent operations jobs. (The number of jobs would depend on the solar technology used, and would likely be similar to the numbers estimated for the similarly sized Gillespie SEZ (i.e., 92 to 1,218 direct construction jobs and 5 to 91 direct operations jobs; least for PV; most for parabolic trough facilities). Adverse impacts could occur due to the need for services for new workers during project construction and operation (e.g., housing, police, firefighters). **Indirect:** The number of jobs would depend on the solar technology used, and would likely be similar to the numbers estimated for the similarly sized Gillespie SEZ (i.e., from 196 to 2,600 indirect construction jobs and 1 to 59 indirect operations jobs). Impacts from project wages and salaries, and tax revenues subsequently circulating through the economy would be minor. **Cumulative:** Cumulative impacts from the presence of large numbers of construction workers could place a short-term strain on local resources. Cumulative impacts during operations would be positive through the creation of additional jobs and income; negative impacts during operations are expected to be small.	See programmatic design features at http://blmsolar.anl.gov/documents/docs/peis/programmatic-design-features/Socioeconomics.pdf and SEZ-specific design features in the RDEP ROD, Table B-4.	On-site mitigation could include requiring developers to secure agreements for local government services as a condition of "Notice to Proceed." See programmatic and SEZ-specific design features.	No. Generally positive impacts expected.

Regional Mitigation Strategy for the Arizona SEZs

Table A-1. (Cont.)

Resource/Issue	Agua Caliente Solar Energy Zone Impacts[1]	On-site Mitigation[2]		Residual Adverse Impacts?[3]
		Avoidance	Minimization	
Soils/Erosion Section 4.2.17	**Direct:** Impacts on soil resources would occur mainly as a result of ground-disturbing activities (e.g., grading, excavating, and drilling), especially during construction. These include removal of topsoil, soil compaction, soil horizon mixing, soil erosion and deposition by wind, soil erosion by water and surface runoff, sedimentation, and soil contamination. Soils in the SEZ are dominated by the Ligurta-Cristobal complex, with 2 to 6 percent slopes, which comprise 1,880 acres, 74% of the SEZ. Areas of desert pavement are present.	Ground disturbance in areas with intact biological soil crusts and desert pavement will be avoided to the extent practicable.	Construction crews should be educated to stay on designated roads and minimize the construction of new roads to minimize soil disturbance and compaction.	Yes. Soils represent a basic component of the ecosystem. Solar development on the SEZ is expected to result in a residual loss of sensitive soils and soil functions.
	According to the Sonoran Desert REA, the majority of the soils on the SEZ have high potential for wind erosion. Therefore, increased wind erosion is likely if grading is needed.	See programmatic design features at http://blmsolar.anl.gov/documents/docs/peis/programmatic-design-features/Soil_Geologic_Hazards.pdf and SEZ-specific design features in the RDEP ROD, Table B-4.	See programmatic and SEZ-specific design features.	
	Indirect: Disturbance of soil can lead to introduction of invasive species and impediments to native seed germination.			
	Cumulative: Cumulative impacts would occur from the disturbance of several renewable energy projects, connecting linear facilities, and other projects in the vicinity of the SEZ, but would be limited through application of design features.			

Table A-1. (Cont.)

Resource/Issue	Agua Caliente Solar Energy Zone Impacts[1]	On-site Mitigation[2]		Residual Adverse Impacts?[3]
		Avoidance	Minimization	
Specially Designated Areas and Lands with Wilderness Characteristics Section 4.2.18	**Direct:** Specially designated areas (SDAs) within 25 miles (40 km) of the SEZ could be visually impacted by solar development. Depending on the solar technology, moderate to strong visual contrasts could be experienced within the Yuma East Undeveloped Special Recreation Management Area (SRMA) 1.1 mi (1.8 km) northwest of the SEZ, within the Gila River Valley Undeveloped SRMA 1.7 mi (2.7 km) south, along the Juan Batista de Anza National Historic Trail and in the Sears Point Core portion of the Sears Point ACEC, both areas within about 5 mi (8 km) south and southeast of the SEZ, and in the Gila River Terraces and Lower Gila Historic Trails ACEC (13 mi to the east of the SEZ). Impacts could include adverse visual effects on the viewshed (including impacts on the night sky viewing) and potential fragmentation of biologically linked areas. There are 140 acres with wilderness characteristics not managed to maintain those characteristics within the SEZ. The SRMS recommends that these areas be non-development areas to avoid impacts to this resource. Solar development would diminish both the nature of these lands and opportunities for solitude and primitive or unconfined recreation to the degree that these characteristics may cease to exist. The result may be a reduction in total acres of lands with wilderness characteristics. **Indirect:** None identified. **Cumulative:** Development of solar facilities and other facilities may result in cumulative effects, particularly visual impacts, on SDAs. **Data Gaps:** Additional analysis may be required to determine if visual impacts could occur in SDAs within the viewshed of the SEZ.	See programmatic design features at http://blmsolar.anl.gov/documents/docs/peis/programmatic-design-features/SDAs_and_LWC.pdf and SEZ-specific design features in the RDEP ROD, Table B-4.	See programmatic and SEZ-specific design features.	Maybe. Residual impacts will be evaluated based on locations of development within the SEZ and project level NEPA.

Table A-1. (Cont.)

Resource/Issue	Agua Caliente Solar Energy Zone Impacts[1]	On-site Mitigation[2] Avoidance	On-site Mitigation[2] Minimization	Residual Adverse Impacts?[3]
Transportation	**Direct:** Palomas Road is approximately 0.5-mile south of the SEZ. Routes in the SEZ are classified as digital linear features or non-motorized routes and usage is documented as light. Impacts are expected to be minor. Development will add traffic to existing roads serving the area. **Indirect:** None identified. **Cumulative:** Cumulative impacts to traffic could occur with multiple developments in the region. **Data Gaps:** Additional data on nearby roads and potential traffic volume during construction/operation is needed.	See programmatic design features at http://blmsolar.anl.gov/documents/docs/peis/programmatic-design-features/Transportation.pdf and SEZ-specific design features in the RDEP ROD, Table B-4.	Local roads would require improvements to accommodate additional traffic. See programmatic and SEZ-specific design features.	No. Through a combination of avoidance, design features, and the establishment of alternative access routes to these areas, the potential impacts can be adequately mitigated onsite.
Visual Section 4.2.22	**Direct:** The Visual Resource Inventory (VRI) value for the SEZ is VRI Class III, indicating moderate visual values. Development will adversely impact visual resources and may impact night skies. However, the visual resource management class is IV and allows for development. **Indirect:** None identified. **Cumulative:** If several projects become visible from one location, or in succession as viewers move through the landscape (such as driving on local roads), these cumulative impacts may make the area less visually appealing.	See programmatic design features at http://blmsolar.anl.gov/documents/docs/peis/programmatic-design-features/Visual.pdf and SEZ-specific design features in the RDEP ROD, Table B-4.	Beyond those required for basic facility and company identification for safety, navigation, and delivery purposes, commercial symbols or signs and associated lighting on buildings and other structures should be prohibited. See programmatic and SEZ-specific design features.	Yes. While on-site mitigation would reduce visual contrasts caused by solar facilities within the SEZ, it would not likely reduce impacts to less than moderate or strong levels for nearby viewers.
Wild Horses and Burros Section 4.2.24	The Agua Caliente SEZ is 17 miles (27 km) or more from any wild horse and burro Herd Management Areas managed by the BLM. Solar energy development within the SEZ would not directly or indirectly affect wild horses and burros.	Not applicable.	Not applicable.	No. The SEZ is not part of a herd management area, and no agency-managed horses or burros are known to exist in the area.

Table A-2. Brenda Solar Energy Zone: Impact Assessment Summary Table

In La Paz County in west-central Arizona, Lake Havasu Field Office —— 1,525 developable acres; up to 305 MW generation capacity, assuming 80% development
Source: Draft and Final Solar PEIS for Brenda SEZ (available at: http://blmsolar.anl.gov/sez/az/brenda/)

Resource/Issue	Brenda Solar Energy Zone Impacts[1]	On-site Mitigation[2]		Residual Adverse Impacts?[3]
		Avoidance	Minimization	
Acoustics Section 8.1.15[4]	**Direct:** Increased noise levels during construction, operation, and decommissioning. **Indirect:** The estimated noise level at the Plomosa Special Recreation Management Area (SRMA) is below the significance threshold. **Cumulative**[5]: If multiple facilities were to be constructed close to the SEZ, residents nearby could be affected by the noise generated, particularly during construction and/or at night when the noise is more discernible due to relatively low background levels. **Data Gaps**[6]: Refined modeling would be warranted along with background noise measurements during project-specific assessments.	Solar facilities must be located far enough away from residences, or include engineering and/or operational methods such that county, state, and/or federal regulations for noise are not exceeded. See programmatic design features at http://blmsolar.anl.gov/documents/docs/peis/programmatic-design-features/Noise.pdf	The hours of daily activities will be limited and noise barriers will be constructed if needed and practicable. Coordination with nearby residents is recommended. See programmatic design features.	Maybe (depends on technology and engineering controls). Generally impacts from solar development are expected to be temporary, localized, and readily mitigated onsite.

[1] The impacts assessment assumed 80% of the SEZ area will be used for solar development.
[2] These columns give examples of avoidance and minimization measures that are specified in the Record of Decision for the Final Solar PEIS and will be required. Additional avoidance and minimization measures proposed by the BLM Interdisciplinary Team are listed and should be evaluated through project-specific environmental analyses. Monitoring is planned to verify the implementation and effectiveness of avoidance and minimization measures.
[3] Residual or unavoidable impacts are residual effects that cannot be adequately mitigated onsite by avoidance and/or minimization.
[4] Section numbers are the same in both the Draft and Final Solar PEIS.
[5] Sections 8.1.22.4 of the Draft and Final Solar PEIS address cumulative impacts, which consider ongoing and reasonably foreseeable activities in the vicinity of the SEZ such as wind, geothermal, mining, agricultural, and commercial development; new roads, traffic, and off-highway vehicle use; and infrastructure such as transmission lines and fences.
[6] Data gaps have not been identified for all resources in this table. Additional data gaps may be identified during future SEZ- or project-specific assessments.

Table A-2. (Cont.)

Resource/Issue	Brenda Solar Energy Zone Impacts[1]	On-site Mitigation[2] Avoidance	On-site Mitigation[2] Minimization	Residual Adverse Impacts?[3]
Air Quality Section 8.1.13	**Direct:** Fugitive dust and equipment exhaust emissions during construction could result in exceedance of Ambient Air Quality Standards (AAQS) for particulate matter (PM) at SEZ boundaries. Specifically, predicted 24-hour PM_{10} and 24-hour and annual $PM_{2.5}$ concentrations could exceed AAQS at the SEZ boundaries and in the immediate surrounding areas during construction of solar facilities. High PM_{10} concentrations would be limited, however, to the immediate areas surrounding the SEZ boundary and would decrease quickly with distance. Generation of fugitive dust may result in exposure to respirable particulates and/or microbes (human health impacts). The majority of the soils on the SEZ have been characterized as having high potential for wind erosion. **Indirect:** Decreased visibility in nearby residential or specially-designated areas due to elevated PM levels from soil disturbance/grading during construction. **Cumulative:** Cumulative effects due to dust emissions would greatest if multiple solar projects had overlapping construction periods. **Data Gaps:** Monitoring for PM during all phases of development will be required to identify levels exceeding AAQS.	See programmatic design features at http://blmsolar.anl.gov/documents/docs/peis/programmatic-design-features/Air_Quality_Climate.pdf	Dust suppression measures will be implemented during all phases of development (construction, operations, and decommissioning) See programmatic design features. Also recommend evaluation of solar panel mounting and other disturbance minimizing technologies in project-level NEPA alternatives (e.g., no grading of the site, retention of maximum native vegetation, use of low emission vehicles, placing gravel on roads, use of "drive and crush" installation). Recommend re-vegetation of the SEZ with native vegetation to increase soil stability as a plan of development feature to further minimize the amount of grading and surface disturbance and promote reduced dust emissions and PM levels.	Maybe (if site is graded). Level of site grading and disturbance to native vegetation would be primary driver of residual impact for full build-out of SEZ. Impacts are not expected to result in noncompliance with National Air Quality Standards.

Table A-2. (Cont.)

Resource/Issue	Brenda Solar Energy Zone Impacts[1]	On-site Mitigation[2]		Residual Adverse Impacts?[3]
		Avoidance	Minimization	
	Direct: Possible impact through loss of carbon storage capacity of the soil (estimated at 100 g carbon/m^2). Preliminary calculations show loss of CO_2 storage capacity as 1.6 tons/acre/yr (4,947 tons/yr for SEZ full build-out), less than 1% of the CO_2 emissions avoided by operation of a solar facility (see below)			
Climate Change Section 5.11.4 of DPEIS for soil storage capacity; 8.1.13 for emissions avoided	**Positive impact:** Solar power generation reduces demand for energy from fossil fuels, and thereby reduces greenhouse gas emissions (from about 509,000–917,000 tons/yr CO_2 avoided at full build-out depending on technology).

Cumulative: Over the long-term the development of solar energy may contribute to reduced greenhouse gas emissions (if the development offsets electricity generation by fossil fuel plants). About 65% of electricity in AZ is produced in fossil fuel plants. Based on data from the Sonoran Desert Rapid Ecoregional Assessment (REA), the SEZ is situated in an area with moderately high to very high potential for future climate change (e.g., increased temperature, decreased precipitation, and changes in vegetation and habitat). | Native vegetation cover and soils will be maintained and grading will be minimized.

See programmatic design features for vegetation at http://blmsolar.anl.gov/documents/docs/peis/programmatic-design-features/Ecological_Resources.pdf | See programmatic design features. | No. Impacts are likely to be positive. No mitigation likely needed. |

Table A-2. (Cont.)

Resource/Issue	Brenda Solar Energy Zone Impacts[1]	On-site Mitigation[2]		Residual Adverse Impacts?[3]
		Avoidance	Minimization	
	Direct: Direct impacts on significant cultural resources could occur in the Brenda SEZ. The SEZ falls within the boundaries of the Desert Training Center/California-Arizona Maneuver Area, which contains scattered resources related to a World War II era training area. Burial locations may be present within or near the Brenda SEZ.		A recently completed archaeological survey has informed the creation of non-development areas within SEZ. An agreement document and a Historic Property Treatment Plan will be written pursuant to Section 106 for the resolution of adverse effects to any historic property included in or eligible for inclusion in the National Register of Historic Places.	Yes. Impacts on non-renewable resources are both irretrievable and irreversible. Tribal consultation may present situations where data recovery or collection onsite is not possible.
Cultural Section 8.1.17	**Indirect:** Erosion impacts on the cultural landscape outside of the SEZ resulting from land disturbances and modified hydrologic patterns; increased accessibility and potential for damage to eligible sites in the non-development area as well as outside of the SEZ. There are several Areas of Critical Environmental Concern (ACECs) in the vicinity of the SEZ that have been determined to be rich in cultural resources. The Harcuvar Mountain West Special Cultural Resource Management Areas is also located 18 mi (29 km) to the northeast. Increased human and vehicle traffic associated with the solar development could impact cultural resources in adjacent or nearby ACECs.	Significant cultural resources clustered in specific areas which retain sufficient integrity will be avoided to the extent possible. See programmatic design features at http://blmsolar.anl.gov/documents/docs/peis/programmatic-design-features/Cultural.pdf	Impacts on culturally significant sites and landscapes in the vicinity of the SEZ at locations such as Ranegras Plain, Granite Wash Pass, Harquahala Mountains, and nearby ACECs and SCRMAs would need to be avoided, minimized, or otherwise mitigated if solar energy development is initiated in the SEZ. See programmatic design features.	Procedures to handle inadvertent discoveries will be addressed in a monitoring and discovery plan developed during the lease process.
	Cumulative: Eligible sites and cultural landscapes could be impacted in the SEZ and adjacent areas. **Data Gaps:** Documentation of a 100% pedestrian archaeological survey of the SEZ is currently being completed. The Section 106 consultation process must also be completed at the project level and has the potential to result in additional information to consider.			

Table A-2. (Cont.)

Resource/Issue	Brenda Solar Energy Zone Impacts[1]	On-site Mitigation[2]		Residual Adverse Impacts?[3]
		Avoidance	Minimization	
Ecology: Vegetation Section 8.1.10	**Direct:** Development will adversely affect characteristic vegetation (e.g., creosote bush, saguaro cactus, paloverde, ironwood, acacia, ocotillo) through destruction and loss of habitat. Development will result in small impacts to the following land types which comprise the SEZ: Creosotebush-White Bursage Desert Scrub and Paloverde-Mixed Cacti Desert Scrub. Sensitive habitats on the SEZ include desert dry wash and dry wash woodland. Development, including vegetation removal, land clearing, grading, and changes in surface water flow may alter soils and vegetation communities and result in the establishment of invasive species and noxious weeds within the SEZ.

Indirect: Loss of native vegetation, increased surface water runoff and related erosion, or through the introduction of invasive species. Establishment of noxious weeds in the SEZ may result in their spreading to adjacent areas.

Cumulative: Solar energy development could be a contributor to cumulative impacts on some vegetation communities, depending on the type, number, and location of other developments in the region. | Dry wash, dry wash woodland, saguaro cactus, and ironwood (including those outside of washes) vegetation communities within the SEZ will be avoided to the extent practicable. A buffer area will be maintained around dry washes and dry wash woodland habitats to reduce the impact potential.

Travel through weed-infested areas will be avoided; vehicles and equipment will be inspected and cleaned to avoid the spread of weeds; ground disturbance will be limited; creation of soil conditions that promote weed germination and establishment will be avoided; seed and plant parts will be disposed of.

See programmatic design features at http://blmsolar.anl.gov/documents/docs/peis/programmatic-design-features/Ecological_Resources.pdf | Appropriate engineering controls will be used to minimize impacts on dry wash, dry wash woodland, and chenopod scrub, including downstream occurrences, resulting from surface water runoff, erosion, sedimentation, altered hydrology, accidental spills, or fugitive dust deposition to these habitats. Appropriate buffers and engineering controls will be determined through agency consultation.

See programmatic design features. | Yes. Development would result in direct removal or disturbance of these native plant communities, special soil environments, and the ecosystem services they provide. |

Regional Mitigation Strategy for the Arizona SEZs

Table A-2. (Cont.)

Resource/Issue	Brenda Solar Energy Zone Impacts[1]	On-site Mitigation[2] Avoidance	On-site Mitigation[2] Minimization	Residual Adverse Impacts?[3]
Ecology: Riparian Areas Section 8.1.10	**Direct:** Development will adversely affect characteristic vegetation (e.g., creosote bush, white bursage, cactus, paloverde, and ironwood) through destruction and loss of habitat. Development, including vegetation removal, land clearing, grading, changes in surface water flow, and dust deposition may alter soils and vegetation communities and result in the establishment of invasive species and noxious weeds within the SEZ. **Indirect:** Loss of native vegetation due to dust deposition from construction and operations, increased surface water runoff and related erosion, or through the introduction of invasive species. Establishment of noxious weeds in the SEZ may result in their spreading to adjacent areas. **Cumulative:** Solar energy development could be a contributor to cumulative impacts on some vegetation communities, depending on the number and location of other developments in the region.	Dry washes, playas, and wetlands within the SEZ will be avoided to the extent practicable. A buffer area will be maintained around wetlands, playas, and dry washes to reduce the potential for impacts. Appropriate engineering controls will be used to minimize impacts on dry wash, dry wash woodland and chenopod scrub, including downstream occurrence, resulting from surface water runoff, erosion, sedimentation, altered hydrology, accidental spills, or fugitive dust deposition to these habitats. Appropriate buffers and engineering controls will be determined through agency consultation. See programmatic design features at http://blmsolar.anl.gov/documents/docs/peis/programmatic-design-features/Ecological_Resources.pdf	See programmatic design features.	Maybe. Depends on the degree of avoidance and engineering controls. Development may alter ephemeral stream channels that can impact flooding and debris flows during storms, groundwater recharge, ecological habitats, and riparian vegetation communities.
Ecology: Invasive & Noxious Weeds Section 8.1.10	**Direct:** Development, including vegetation removal, land clearing, grading, changes in surface water flow, and dust deposition may alter soils and vegetation communities and result in the establishment of invasive species and noxious weeds within the SEZ. **Indirect:** Loss of native vegetation due to dust deposition from construction and operations, increased surface water runoff and related erosion, or through the introduction of invasive species. Establishment of noxious weeds in the SEZ may result in their spreading to adjacent areas. **Cumulative:** Solar energy development could be a contributor to cumulative impacts on some vegetation communities, depending on the number and location of other developments in the region.	Travel through weed infested areas will be avoided; vehicles and equipment will be inspected and cleaned to avoid spread of weeds; ground disturbance will be limited, creation of soil conditions that promote weed germination and establishment will be avoided, and disposal of seed and plant parts will be disposed of to reduce invasive species establishment. See programmatic design features at http://blmsolar.anl.gov/documents/docs/peis/programmatic-design-features/Ecological_Resources.pdf	Impacts will be minimized through development of a Weed Management Plan and use of weed-free seed to support re-vegetation efforts, control invasive species, and prevent increase in fires. See programmatic design features.	Maybe. On-site mitigation will reduce, but not eliminate, the potential for invasive species. The degree of disturbance creates a significant opportunity for the establishment of invasive species and weeds.

Regional Mitigation Strategy for the Arizona SEZs

Table A-2. (Cont.)

Resource/Issue	Brenda Solar Energy Zone Impacts[1]	On-site Mitigation[2]		Residual Adverse Impacts?[3]
		Avoidance	Minimization	
Ecology: Terrestrial Wildlife and Aquatic Biota Section 8.1.11	**Direct:** Loss of habitat and connectivity for several species of amphibians, reptiles, mammals, bats, and invertebrates. Ground disturbance, fugitive dust generated by project activities, lighting, vegetation clearing, spread of invasive species, accidental spills, harassment, and impacts on ephemeral washes could impact wildlife within the SEZ. Impacts from noise on wildlife could occur, especially on bat species, if the SEZ is located near any bat roosts. **Indirect:** Outside the SEZ, impacts could occur from habitat loss or modification, increased human presence in the area, surface runoff, dust, noise, lighting, or accidental spills. **Cumulative:** Cumulative effects on some species could rise to a level of moderate, given the large acreages potentially disturbed and depending on the type, number, and location of other developments in the region. **Data Gaps:** Impacts on terrestrial wildlife from construction noise would have to be considered on a project-specific basis, especially for bat species.	Wetlands, washes, and riparian areas identified during site-specific surveys will be avoided. See programmatic design features at http://blmsolar.anl.gov/documents/docs/peis/programmatic-design-features/Ecological_Resources.pdf	The fencing around the solar energy development should not block the free movement of mammals, particularly big game species. Appropriate engineering controls will be implemented to minimize the amount of contaminants and sediment entering Bouse Wash. See programmatic design features.	Yes. Development of the Brenda SEZ will likely impact up to 3,348 acres of wildlife habitat. Level of site grading and disturbance to native vegetation would be primary driver of residual impact for full build-out of SEZ.

Table A-2. (Cont.)

Resource/Issue	Brenda Solar Energy Zone Impacts[1]	On-site Mitigation[2]		Residual Adverse Impacts?[3]
		Avoidance	Minimization	
Ecology: Migratory Birds Section 8.1.11.2	**Direct:** Loss of individuals, habitat, and connectivity for several species protected under the Migratory Bird Treaty Act. Noise, lighting, and vegetation clearing could impact migratory birds using the SEZ. There is potential for water birds to be attracted to solar fields (because they look like water) and collide with solar panels. Burning of wings in the solar radiation field between heliostats and power towers has been observed. There may also be impacts on night sky that may alter bird migratory behavior and habitat use. Priority migratory bird species that may occur on or near the SEZ include Gila woodpecker and gilded flicker.[21] **Indirect:** Outside the SEZ, impacts could occur from habitat loss. **Cumulative:** Impacts to migratory birds could occur; depending on the number and location of other developments in the region. **Data Gaps:** Additional research needed on solar development impacts on migratory birds, Impacts on migratory birds from construction noise would have to be considered on a project-specific basis.	See programmatic design features at http://blmsolar.anl.gov/documents/docs/peis/programmatic-design-features/Ecological_Resources.pdf Effects to individual migratory birds and bird nests can be avoided by not constructing during the breeding season. Timing limitation should be enforced from May 15—July 15 for any surface disturbing activities to protect migratory bird nesting and brood rearing, If construction takes place during the breeding season, nest surveys will be conducted.	See programmatic design features. Also recommend implementation of technologies that minimize the amount of reflective surfaces, or alter how the surfaces are perceived by wildlife, that will reduce the "lake effect" in attracting migratory birds and other wildlife.	Yes. Development of the Brenda SEZ will likely impact up to 3,348 acres of migratory bird habitat. Some level of bird injury/fatality has been observed for all types of solar facilities (through collisions with equipment or from burns). Research is ongoing to quantify impacts and identify effective mitigation measures.

[21] Priority migratory bird species for the SEZ were determined based on those species discussed in the Lake Havasu RMP, the distribution of Arizona Natural Heritage Program tracked species, and USFWS Birds of Conservation Concern in the Arizona Habimap tool (http://www.habimap.org/).

Table A-2. (Cont.)

| Resource/Issue | Brenda Solar Energy Zone Impacts[1] | On-site Mitigation[2] | | Residual Adverse Impacts?[3] |
		Avoidance	Minimization	
Ecology: Plant Special Status Species Section 8.1.12	**Direct:** No Endangered Species Act (ESA)-listed or BLM-listed sensitive plant species have been identified that have suitable habitat within the SEZ. Ground disturbance, land clearing and grading, fugitive dust generated by project activities, and the spread of invasive species may result in loss of special status plant species habitat, if present, and might result in loss of individual plants. **Indirect:** Indirect impacts to individuals and habitat could occur from surface runoff, dust, or accidental spills. No ESA- or BLM-listed sensitive plant species have been identified that have suitable habitat on or near the SEZ. Potential impacts from groundwater withdrawals. **Cumulative:** There would be no cumulative impacts on special status plant species unless they are discovered during pre-disturbance surveys (cumulative impacts then might be due to habitat destruction and overall development and fragmentation of the area). **Data Gaps:** Although habitat for listed species has not been identified within the SEZ, pre-disturbance surveys are required to identify the presence and abundance of special status species.	Based on data from pre-disturbance surveys, disturbance to occupied habitats would be avoided to the extent practicable. See programmatic design features at http://blmsolar.anl.gov/documents/docs/peis/programmatic-design-features/Ecological_Resources.pdf	If avoidance is not possible for some species, translocation of individuals from areas of direct effects or compensatory mitigation may be employed. See programmatic design features.	No, unless special status plant species are discovered during pre-disturbance surveys. There are no known SSS plant species within the SEZ.

Regional Mitigation Strategy for the Arizona SEZs

Table A-2. (Cont.)

Resource/Issue	Brenda Solar Energy Zone Impacts[1]	On-site Mitigation[2] Avoidance	Minimization	Residual Adverse Impacts?[3]
Ecology: Animal Special Status Species Section 8.1.12	**Direct:** Ground disturbance, land clearing and grading, and fugitive dust generated by project activities would result in loss of special status animal species habitat, if present, and might result in loss of individual animals. Impacts from noise on special status wildlife could also occur. Solar PEIS analyses indicated that development on the SEZ could directly disturb individuals or habitat for one candidate species for listing under the ESA (i.e., **Sonoran desert tortoise** [22]) and seven BLM-sensitive special status animal species (lowland leopard frog, desert rosy boa, American peregrine falcon, Sonoran bald eagle, Western burrowing owl, California leaf-nosed bat, and Townsend's big-eared bat,). Subsequent BLM analyses show that Brenda SEZ is actually outside the occurrence area of the following species: lowland leopard frog, desert rosy boa, American peregrine falcon, and Sonoran Bald eagle. Impacts to these species will not be further considered. No Category 1, 2, or 3 desert tortoise habitat has been identified within the SEZ; however, Category 2 desert tortoise habitat occurs outside the SEZ to the south and west. Desert tortoises may still occur in lower quality habitat on the SEZ where they may be directly impacted by solar development. **Indirect:** Indirect impacts to individuals and animal habitat outside of the SEZ could occur due to surface runoff, dust, noise, lighting, or accidental spills. Suitable habitat for 2 BLM-sensitive animal species occurs within 5 mi (8 km) of the SEZ boundary. However, impacts would be small, with losses of less than 1 percent of these species' habitat in the region. **Cumulative:** There could be cumulative impacts on some special status animal species due to habitat destruction and overall development and fragmentation of the area. **Data Gaps:** Pre-disturbance surveys are required to identify the presence and abundance of special status species.	Compliance with the Bald & Golden Eagle Protection Act would be ensured and Eagle Take Guidance would be followed (if necessary). Based on data from pre-disturbance surveys, disturbance to suitable habitats would be avoided to the extent practicable. See programmatic design features at http://blmsolar.anl.gov/documents/docs/peis/programmatic-design-features/Ecological_Resources.pdf	If avoidance is not possible for some species, translocation of individuals from areas of direct effects or compensatory mitigation may be employed. Regarding on-site avoidance and minimization, consultation with the USFWS will be conducted to address the potential for impacts on ESA-listed and proposed species and to identify mitigation measures for implementation. See programmatic design features.	Yes. Animal SSS along with other wildlife represent a basic component of the ecosystem. Level of site grading and disturbance to native vegetation would be the primary driver of residual impact to functional habitat for full build out of the SEZ.

---

[22] Species in bold text have been recorded 5 miles (8 km) of the SEZ.

Regional Mitigation Strategy for the Arizona SEZs

Table A-2. (Cont.)

Resource/Issue	Brenda Solar Energy Zone Impacts[1]	On-site Mitigation[2]		Residual Adverse Impacts?[3]
		Avoidance	Minimization	
Environmental Justice Section 8.1.20	**Direct:** There is a minority population within a 50-mile (80 km) radius of the SEZ, so any adverse impacts of solar projects could affect this population. There are no low-income populations within a 50-mile radius of the SEZ. **Indirect:** None identified. **Cumulative:** Contributions from solar development in the SEZ would likely be small and would not be expected to significantly contribute to cumulative impacts on minority populations within the 50-mile geographic extent of effects.	See programmatic design features at http://blmsolar.anl.gov/documents/docs/peis/programmatic-design-features/Environmental_Justice.pdf	See programmatic design features.	Maybe.
Hydrology: Surface Water Section 8.1.9	**Direct:** Land clearing, land leveling, vegetation removal, and spills and runoff associated with development of the SEZ could increase surface runoff, reduce infiltration/recharge, cause loss of ephemeral stream networks, cause a reduction in evapotranspiration rates, increase sediment transport (by water), change sediment transport (by wind), and degrade water quality. There are no perennial surface water features, flood hazards, or wetlands within Brenda SEZ. **Indirect:** Indirect impacts from development and groundwater use on ephemeral and perennial surface water features could occur. **Cumulative:** Alterations to ephemeral stream networks can alter groundwater recharge and surface runoff processes potentially impacting the basin-scale water balance and water quality aspects of water features receiving surface runoff.	See programmatic design features at http://blmsolar.anl.gov/documents/docs/peis/programmatic-design-features/Water.pdf	See programmatic design features.	Yes. Hydrology is a basic component of the ecosystem. Reconfiguration of topography for solar development would have residual impacts to surface hydrology with potential impacts on other resources, including vegetation.

A-31

Regional Mitigation Strategy for the Arizona SEZs

Table A-2. (Cont.)

Resource/Issue	Brenda Solar Energy Zone Impacts[1]	On-site Mitigation[2]		Residual Adverse Impacts?[3]
		Avoidance	Minimization	
Hydrology: Water Quality and Groundwater Availability Section 8.1.9	**Direct:** Groundwater withdrawals for development may cause declines in groundwater elevations that can impact water availability for surface water features, vegetation, ecological habitats, regional groundwater flow paths, and other groundwater users in the basin. The SEZ is located in the Ranegras Plain groundwater basin where available groundwater occurs primarily in basin-fill deposits. **Indirect:** Groundwater withdrawals for solar energy facilities may affect other groundwater users in the basin. **Cumulative:** Cumulative impacts on groundwater could occur when combined with other future developments in the region.	Groundwater analyses suggest that full build-out of wet-cooled technologies is not feasible. See programmatic design features at http://blmsolar.anl.gov/documents/ docs/peis/programmatic-design-features/Water.pdf	The SEZ is located in Water Protection Zone 3 and new water uses and withdrawals are restricted to panel washing and sanitary uses only.[23] For mixed-technology development scenarios, any proposed wet-cooled projects would be required to retire existing groundwater uses and utilize water conservation practices. See programmatic design features.	No. It is possible for impacts on groundwater aquifers to be avoided or minimized.
Lands & Realty Section 8.1.2	**Direct:** Development of the SEZ could disturb 3,348 acres (13.5 km²). There is a small portion of a ROW for a fiber-optic line that parallels Highway 60 that overlaps the SEZ. **Indirect:** Increased traffic and increased access to previously remote areas also could change the overall character of the landscape. **Cumulative:** Projects within the SEZ would make only a small contribution to cumulative impacts because of its relatively small size.	See programmatic design features at http://blmsolar.anl.gov/documents/ docs/peis/programmatic-design-features/Lands_and_Realty.pdf	Any potential impacts on the existing county road should be discussed with the county. See programmatic design features.	No. By regulation, any new activity must occur in deference to existing rights. Thus, potential impacts have been avoided.

---

[23] Unavoidable adverse impacts are possible if groundwater is used. However, wet cooling was not considered a feasible option in the Solar PEIS ROD. Additional restrictions identified in the RDEP ROD for Water Protection Zone 3 would further limit the potential for residual impacts to occur.

Table A-2. (Cont.)

Resource/Issue	Brenda Solar Energy Zone Impacts[1]	On-site Mitigation[2]		Residual Adverse Impacts?[3]
		Avoidance	Minimization	
Livestock Grazing Section 8.3.4.1	**Direct:** SEZ is located within the Crowder-Weisser Grazing Allotment; the land within the SEZ constitutes less than 2 percent of the allotment. Due to the large size of the allotment, it might be possible to accommodate any lost animal unit months elsewhere in the allotment. If that is not possible, there would be an undetermined adverse economic impact upon the permittee. **Indirect:** None identified. **Cumulative:** Other development in the area of the SEZ could result in cumulative impacts on grazing.	See programmatic design features at http://blmsolar.anl.gov/documents/docs/peis/programmatic-design-features/Rangeland_Resources.pdf	See programmatic design features.	No.
Military & Civilian Aviation Section 8.1.6	**Direct:** The SEZ is located within an extensive web of military training routes (MTRs), and the entire SEZ is covered by a combination of three MTRs with 300-foot (91-meter) above-ground-level operating limits. The military has said that solar or transmission facilities in excess of 250 feet (76 meters) tall would adversely affect the use of the MTRs. The Blythe Airport is about 48 miles (77 km) west of the SEZ, and the Parker airport (Avi Suquilla Airport) is about 38 miles (61 km) northwest of the SEZ. Neither of these airports has regularly scheduled passenger or freight service. **Indirect:** None identified. **Cumulative:** Solar development occurring throughout the region, which is currently largely undeveloped, could result in small cumulative effects on the system of MTRs. Such effects would be limited by mitigations developed in consultation with the military.	See programmatic design features at http://blmsolar.anl.gov/documents/docs/peis/programmatic-design-features/Military_Civilian_Aviation.pdf	Coordination with the military will be required on a project-specific basis to ensure that solar facilities do not interfere with operations. See programmatic design features.	Maybe (with respect to MTRs). Residual impacts will be evaluated based on coordination with the military and project-level NEPA.
Minerals Section 8.1.8 and Section 8.1.24 of the Final PEIS	**Direct:** There are no locatable mining claims within the SEZ. The SEZ has been withdrawn from mineral entry for a period of 20 years, precluding impacts from many types of mining activities. **Indirect:** None identified. **Cumulative:** None identified.	See programmatic design features at http://blmsolar.anl.gov/documents/docs/peis/programmatic-design-features/Mineral_Resources.pdf	See programmatic design features.	No

Regional Mitigation Strategy for the Arizona SEZs

Table A-2. (Cont.)

Resource/Issue	Brenda Solar Energy Zone Impacts[1]	On-site Mitigation[2]		Residual Adverse Impacts?[3]
		Avoidance	Minimization	
Native American Concerns Section 8.1.18	**Direct:** A tribe has indicated that some of the land in the SEZ lies within its tribal traditional use area. The tribe has expressed concerns regarding the loss of many resources, including natural habitat, wild plant resources, game animals, viewsheds, and cremation or burial sites. As consultations continue, it is possible that other Native American concerns regarding solar energy development within the SEZ will emerge. Removal of cultural resources is a concern to the tribes.	Known human burial sites and rock art (panels of petroglyphs and/or pictographs) will be avoided. The BLM will consult with Indian tribes regarding the potential for unanticipated human remains and associated cultural items (as defined under the Native American Graves Protection and Repatriation Act), before a solar project is authorized. The purpose will be to discuss general guidance on treatment of cultural items.		
	Indirect: General habitat loss with vegetation clearing and water reduction that could affect species and ecosystem health.	Springs and other water sources that are or may be sacred or culturally important, culturally important plant and wildlife species, and visual intrusion on sacred sites will be avoided to the extent practicable.	See programmatic design features.	Yes. Consultation on project applications will determine whether regional mitigation for Native American Concerns is warranted.
	Cumulative: Development of solar energy facilities in combination with the development of other planned and foreseeable projects in the area would likely reduce the traditionally important plant and animal resources available to the tribes. Although some of these plant species are abundant, any level of impact may be of concern for the tribes.	EO13007 requires executive branch agencies to accommodate access to and ceremonial use of Indian sacred sites and to avoid adverse impacts to the physical integrity of such a site. Because solar facilities will be fenced and security procedures will limit or eliminate access, if a sacred site was declared, it may not be possible to mitigate impacts, other than through avoidance.	Mitigate onsite to comply with EO13007. BLM could facilitate the harvest of creosote prior to ground disturbance for instance, if identified as a concern.	
	Data Gaps: Documentation of an archaeological survey of the entire SEZ is currently being completed and results will be shared with the tribes. Government-to-Government consultation for projects will be required to determine issues of Native American concern.	See programmatic design features at http://blmsolar.anl.gov/documents/docs/peis/programmatic-design-features/Native_American_Concerns.pdf		

Regional Mitigation Strategy for the Arizona SEZs

Table A-2. (Cont.)

Resource/Issue	Brenda Solar Energy Zone Impacts[1]	On-site Mitigation[2]		Residual Adverse Impacts?[3]
		Avoidance	Minimization	
Paleontology Section 8.1.16	**Direct:** The SEZ is located in an area classified as Potential Fossil Yield Classification (PFYC) Class 3b. It has a low to undetermined potential for paleontological resources. **Indirect:** None identified. **Cumulative:** Cumulative impacts are dependent on whether significant resources are found within the SEZ and in additional project areas in the region. **Data Gaps:** Potential for impacts is unknown. A more detailed assessment of the geological deposits of the SEZ is needed to determine whether a paleontological survey is warranted for a specific project.	See programmatic design features at http://blmsolar.anl.gov/documents/docs/peis/programmatic-design-features/Paleo.pdf	The BLM will be notified immediately upon discovery of fossils. Work will be halted at the fossil site and continued elsewhere until qualified personnel, such as a paleontologist, can visit the site. He/she will determine if the site is significant and make recommendations for collection or other resource protection, if warranted. See programmatic design features.	No. Design features will reduce the risk that any paleontological resources that are discovered will be destroyed.
Public Access and Recreation Section 8.1.5	**Direct:** Dispersed recreational users would be displaced from areas developed for solar energy production within the Brenda SEZ. **Indirect:** Indirect effects on recreation use would occur on lands near the solar facilities, primarily the Plomosa SRMA (0.9 mi [1.5 km] from the SEZ), and would result from the change in the overall character of undeveloped BLM-administered lands to an industrialized, developed area. La Posa Destination SRMA and Yuma East Undeveloped SRMA are also within 25 mi (40 km) of the SEZ. People seeking more rural or primitive surroundings for recreation may experience a reduction in recreational opportunities and/or a degraded recreational experience. Privately owned Recreational Vehicle parks may be impacted due to limited recreation areas. **Cumulative:** Multiple developments in the vicinity of the SEZ could cumulatively reduce recreational opportunities.	See programmatic design features at http://blmsolar.anl.gov/documents/docs/peis/programmatic-design-features/Public_Access_and_Recreation.pdf	See programmatic design features.	Yes.

Regional Mitigation Strategy for the Arizona SEZs

Table A-2. (Cont.)

| Resource/Issue | Brenda Solar Energy Zone Impacts[1] | On-site Mitigation[2] | | Residual Adverse Impacts?[3] |
		Avoidance	Minimization	
Socioeconomics Section 8.1.19	**Direct:** Impacts to local economy as a result of expenditures of wages and salaries and the collection of state sales and income taxes. From 118 to 1,557 direct construction jobs and 6 to 117 direct operations jobs could be created (least for PV; most for parabolic trough facilities). Adverse impacts could occur due to the need for services for new workers during project construction and operation (e.g., housing, police, firefighters). **Indirect:** From 236 to 3,126 indirect construction jobs and 2 to 74 indirect operations jobs could be created. Impacts from project wages and salaries, and tax revenues subsequently circulating through the economy would be minor. **Cumulative:** Impacts overall would be positive, through the creation of additional jobs and income. The negative impacts, including some short-term disruption of rural community quality of life, are expected to be small.	See programmatic design features at http://blmsolar.anl.gov/documents/docs/peis/programmatic-design-features/Socioeconomics.pdf	See programmatic design features. Additionally, on-site mitigation could include requiring developers to secure agreements for local government services as a condition of "Notice to Proceed".	No. Generally positive impacts expected.

Regional Mitigation Strategy for the Arizona SEZs

Table A-2. (Cont.)

Resource/Issue	Brenda Solar Energy Zone Impacts[1]	On-site Mitigation[2]		Residual Adverse Impacts?[3]
		Avoidance	Minimization	
Soils/Erosion Section 8.1.7	**Direct:** Impacts on soil resources would occur mainly as a result of ground-disturbing activities (e.g., grading, excavating, and drilling), especially during construction of a solar project. These include removal of topsoil, soil compaction, soil horizon mixing, soil erosion and deposition by wind, soil erosion by water and surface runoff, sedimentation, and soil contamination. Soils within the SEZ are predominantly the loams and sandy loams of soil series Pahaka-Estrella-Antho. Because of their fine-grained texture, they are moderately susceptible to wind erosion. Soil contamination from spills could occur. Based on an evaluation of data in the Sonoran Desert REA, the majority of the soils on the SEZ have high potential for wind erosion. Therefore, increased wind erosion is likely if grading occurs. **Indirect:** Disturbance of soil can lead to introduction of invasive species. **Cumulative:** Cumulative impacts would occur from the disturbance of several renewable energy projects, connecting linear facilities, and other projects in the vicinity of the SEZ, but would be limited through application of design features.	See programmatic design features at http://blmsolar.anl.gov/documents/docs/peis/programmatic-design-features/Soil_Geologic_Hazards.pdf	Construction crews should be educated to stay on designated roads and minimize the construction of new roads to minimize soil disturbance and compaction. See programmatic design features.	Yes. Soils represent a basic component of the ecosystem. Solar development on the SEZ is expected to result in a residual loss of sensitive soils and soil functions.
Specially Designated Areas and Lands with Wilderness Characteristics Section 8.1.3	**Direct:** Specially designated areas (SDAs) within 25 miles of the SEZ that could be impacted by solar development are East Cactus Plain Wilderness Area (WA), Kofa WA, New Water Mountain WA, Cactus Plain Wilderness Study Area (WSA), Dripping Springs ACEC, Harquahala ACEC, and Kofa National Wildlife Refuge (visual impacts estimated as minimal for all). Impacts could include adverse visual effects on the viewshed (including impacts on night sky viewing). There are no undesignated areas with wilderness characteristics near the SEZ. **Indirect:** None identified. **Cumulative:** Development of solar facilities and other facilities may result in cumulative effects, particularly visual impacts, on SDAs.	See programmatic design features at http://blmsolar.anl.gov/documents/docs/peis/programmatic-design-features/SDAs_and_LWC.pdf	See programmatic design features.	Yes. Residual impacts will be evaluated based on locations of development within the SEZ and project level NEPA.

Table A-2. (Cont.)

Resource/Issue	Brenda Solar Energy Zone Impacts[1]	On-site Mitigation[2]		Residual Adverse Impacts?[3]
		Avoidance	Minimization	
Transportation Section 8.1.21	**Direct:** Development will add traffic to existing roads serving the area. The volume of traffic on U.S. 60 could represent an increase in traffic of about 130 percent during construction. Such traffic levels would represent an increase in the traffic levels experienced on I-10 or State Route 72 at their junctions with U.S. 60. Local roads would also be impacted. **Indirect:** None identified. **Cumulative:** Cumulative impacts to traffic could occur with multiple developments in the region.	See programmatic design features at http://blmsolar.anl.gov/documents/docs/peis/programmatic-design-features/Transportation.pdf	Local roads would require improvements to accommodate additional traffic. See programmatic design features.	No. Through a combination of avoidance, design features, and the establishment of alternative access routes to these areas, the potential impacts can be adequately mitigated onsite.
Visual Section 8.1.14	**Direct:** The Visual Resource Inventory (VRI) value for the SEZ and immediate surroundings are VRI Class IV, indicating low visual values. Development will adversely impact visual resources and may impact night skies. The Solar PEIS identified moderate to strong visual contrasts for the Plomosa SRMA, La Posa Destination SRMA, U.S. Highway 60, Interstate 10, and the towns of Vicksburg and Brenda. **Indirect:** None identified. **Cumulative:** If several projects become visible from one location, or in succession as viewers move through the landscape (such as driving on local roads), these cumulative impacts may make the area less visually appealing.	See programmatic design features at http://blmsolar.anl.gov/documents/docs/peis/programmatic-design-features/Visual.pdf	Beyond those required for basic facility and company identification for safety, navigation, and delivery purposes, commercial symbols or signs and associated lighting on buildings and other structures should be prohibited. See programmatic design features.	Yes. While on-site mitigation would reduce visual contrasts caused by solar facilities within the SEZ, it would not likely reduce impacts to less than moderate or strong levels for nearby viewers.
Wild Horses and Burros Section 8.1.4.2	The Brenda SEZ is 19 miles (31 km) or more from any wild horse and burro Herd Management Areas managed by the BLM and more than 50 mi (80 km) from any wild horse and burro territory administered by the U.S. Forest Service. Solar energy development within the SEZ would not directly or indirectly affect wild horses and burros that are managed by these agencies.	Not applicable.	Not applicable.	No. The SEZ is not part of a herd management area, and no agency-managed horses or burros are known to exist in the area.

Table A-3. Gillespie Solar Energy Zone: Impact Assessment Summary Table

In Maricopa County in west-central Arizona, Lower Sonoran Field Office — 1,785 developable acres, up to 357 MW generation capacity, assuming 80% development

Source: Draft and Final Solar PEIS for Gillespie SEZ available at: (http://blmsolar.anl.gov/sez/az/gillespie/)

Resource/Issue	Gillespie Solar Energy Zone Impacts[1]	On-site Mitigation[2]		Residual Adverse Impacts[3]? (include justification)
		Avoidance	Minimization	
Acoustics Section 8.3.15[4]	**Direct:** Increased noise levels during construction, operation, and decommissioning could cause impacts, particularly for activities occurring near the southeastern boundary of the SEZ, close to the nearest residences. Estimated noise levels at the nearest residences would not exceed EPA's guideline level. **Indirect:** Noise from solar development in the SEZ is not likely to adversely affect any of the nearby specially designated areas. **Cumulative[5]:** If multiple facilities were to be constructed close to the SEZ, residents nearby could be affected by the noise generated, particularly during construction and/or at night when the noise is more discernible due to relatively low background levels. **Data Gaps[6]:** Refined modeling would be warranted along with background noise measurements during project-specific assessments.	Solar facilities must be located far enough away from residences, or include engineering and/or operational methods such that county, state, and/or federal regulations for noise are not exceeded. See programmatic design features at http://blmsolar.anl.gov/documents/docs/peis/programmatic-design-features/Noise.pdf	The hours of daily activities will be limited and noise barriers will be constructed if needed and practicable. Coordination with nearby residents is recommended. See programmatic design features.	Maybe (depends on technology and engineering controls). Generally impacts from solar development are expected to be temporary, localized, and readily mitigated onsite.

[1] The impacts assessment assumed 80% of the SEZ area will be used for solar development.

[2] These columns give examples of avoidance and minimization measures that are specified in the Record of Decision for the Final Solar PEIS and will be required. Additional avoidance and minimization measures proposed by the BLM Interdisciplinary Team are listed and should be evaluated through project-specific environmental analyses. Monitoring is planned to verify the implementation and effectiveness of avoidance and minimization measures.

[3] Residual or unavoidable impacts are residual effects that cannot be adequately mitigated onsite by avoidance and/or minimization.

[4] Section numbers are the same in both the Draft and Final Solar PEIS.

[5] Sections 8.3.22.4 of the Draft and Final Solar PEIS address cumulative impacts, which consider ongoing and reasonably foreseeable activities in the vicinity of the SEZ such as wind, geothermal, mining, agricultural, and commercial development; new roads, traffic, and off-highway vehicle use; and infrastructure such as transmission lines and fences.

[6] Data gaps have not been identified for all resources in this table. Additional data gaps may be identified during future SEZ- or project-specific assessments.

Regional Mitigation Strategy for the Arizona SEZs

Table A-3. (Cont.)

Resource/Issue	Gillespie Solar Energy Zone Impacts[1]	On-site Mitigation[2] Avoidance	On-site Mitigation[2] Minimization	Residual Adverse Impacts[3]? (include justification)
Air Quality Section 8.3.13	**Direct:** Fugitive dust and equipment exhaust emissions during construction could result in exceedance of Ambient Air Quality Standards (AAQS) for particulate matter (PM) at SEZ boundaries. However, some existing background PM levels already exceed the standards. Specifically, predicted 24-hour PM_{10} and 24-hour and annual $PM_{2.5}$ concentrations could exceed AAQS at the SEZ boundaries and in the immediate surrounding areas during the construction of solar facilities. High PM_{10} concentrations would be limited, however, to the immediate areas surrounding the SEZ boundary and would decrease quickly with distance. Generation of fugitive dust may result in exposure to respirable particulates and/or microbes (human health impacts). The majority of the soils on the SEZ have been characterized as having high potential for wind erosion. **Indirect:** Decreased visibility in nearby residential or specially-designated areas due to elevated PM levels from soil disturbance/grading during construction. **Cumulative:** Cumulative effects due to dust emissions would be greatest if multiple solar projects had overlapping construction periods. **Data Gaps:** Monitoring for PM during construction, operation, and decommissioning will be required to identify levels exceeding AAQS.	See programmatic design features at http://blmsolar.anl.gov/documents/docs/peis/programmatic-design-features/Air_Quality_Climate.pdf	Dust suppression measures will be implemented during all phases of development (construction, operations, and decommissioning). See programmatic design features. Also recommend evaluation of solar panel mounting and other disturbance minimizing technologies in project-level NEPA alternatives (e.g., no grading of the site, retention of maximum native vegetation, use of low emission vehicles, placing gravel on roads, use of "drive and crush" installation). Recommend re-vegetation of the SEZ with native vegetation to increase soil stability as a plan of development feature to further minimize the amount of grading and surface disturbance and promote reduced dust emissions and PM levels.	Maybe (if site is graded). Level of site grading and disturbance to native vegetation would be primary driver of residual impact for full build-out of SEZ. Impacts are not expected to result in noncompliance with National Air Quality Standards.

Regional Mitigation Strategy for the Arizona SEZs

Table A-3. (Cont.)

Resource/Issue	Gillespie Solar Energy Zone Impacts[1]	On-site Mitigation[2]		Residual Adverse Impacts[3]?
		Avoidance	Minimization	(include justification)
Climate Change Section 5.11.4 of DPEIS for soil storage capacity; 8.3.13 for emissions avoided	**Direct:** Possible impact through loss of carbon storage capacity of the soil (estimated at 100 g carbon/m^2). Preliminary calculations show loss of CO_2 storage capacity as 1.6 tons/acre/yr (3,351 tons/yr for SEZ full build-out), less than 1 percent of the CO_2 emissions avoided by operation of a solar facility (see below). **Positive impact:** Solar power generation reduces demand for energy from fossil fuels, and thereby reduces greenhouse gas emissions (from about 347,000–624,000 tons/yr CO_2 avoided at full build-out depending on technology). **Cumulative:** Over the long-term, the development of solar energy may contribute to reduced greenhouse gas emissions (if the development offsets electricity generation by fossil fuel plants. About 65% of electricity in AZ is produced in fossil fuel plants. Based on data from the Sonoran Desert Rapid Ecoregional Assessment (REA), the SEZ is situated in an area with moderate to moderately low potential for future climate change (e.g., increased temperature, decreased precipitation, and changes in vegetation and habitat).	Native vegetation cover and soils will be maintained and grading will be minimized. See programmatic design features for vegetation at http://blmsolar.anl.gov/documents/docs/peis/programmatic-design-features/Ecological_Resources.pdf	See programmatic design features.	No. Impacts are likely to be positive. No mitigation likely needed.

Regional Mitigation Strategy for the Arizona SEZs

Table A-3. (Cont.)

Resource/Issue	Gillespie Solar Energy Zone Impacts[1]	On-site Mitigation[2] Avoidance	On-site Mitigation[2] Minimization	Residual Adverse Impacts[3]? (include justification)
	Direct: Development may adversely affect cultural resources. **Indirect:** Erosion impacts on the cultural landscape outside of the SEZ resulting from land disturbances and modified hydrologic patterns; increased accessibility and potential for damage to eligible sites outside of the SEZ (if present).	Significant resources clustered in specific areas which retain sufficient integrity will be avoided to the extent possible.	Recordation of historic structures through Historic American Building Survey/Historic American Engineering Record protocols through the National Park Service would be appropriate and could be required if any historic structures or features would be affected, or if the Gillespie Dam Highway Bridge (i.e., Old U.S. 80 bridge) were used as part of an off-site access route.	Maybe, pending review of the archaeological survey documentation. Impacts on non-renewable resources are both irretrievable and irreversible. Tribal consultation may present situations where data recovery or collection onsite is not possible.
Cultural Section 8.3.17	**Cumulative:** Dependent on whether eligible sites or landscapes are present and impacted in the SEZ and adjacent areas. **Data Gaps:** Documentation of a 100% pedestrian archaeological survey of the SEZ is currently being completed. The Section 106 consultation process must also be completed at the project level and has the potential to result in additional information to consider.	See programmatic design features at http://blmsolar.anl.gov/documents/docs/peis/programmatic-design-features/Cultural.pdf	The archaeological survey has informed the creation of non-development areas within the SEZ. An agreement document and a Historic Property Treatment Plan will be written pursuant to Section 106 for the resolution of adverse effects to any historic property included in or eligible for inclusion in the National Register of Historic Places. See programmatic design features.	Procedures to handle inadvertent discoveries will be addressed in a monitoring and discovery plan developed during the lease process.

Table A-3. (Cont.)

Resource/Issue	Gillespie Solar Energy Zone Impacts[1]	On-site Mitigation[2]		Residual Adverse Impacts[3]? (include justification)
		Avoidance	Minimization	
Ecology: Vegetation Section 8.3.10	**Direct:** Development will adversely affect characteristic vegetation (e.g., creosote bush, white bursage, cactus, paloverde, and ironwood) through destruction and loss of habitat. Development will result in small impacts to the following land types which comprise the SEZ: Creosotebush-White Bursage Desert Scrub and Paloverde-Mixed Cacti Desert Scrub. Development, including vegetation removal, land clearing, grading, changes in surface water flow, and dust deposition may alter soils and vegetation communities and result in the establishment of invasive species and noxious weeds within the SEZ. **Indirect:** Loss of native vegetation due to dust deposition from construction and operations, increased surface water runoff and related erosion, or through the introduction of invasive species. Establishment of noxious weeds in the SEZ may result in their spreading to adjacent areas. **Cumulative:** Solar energy development could be a contributor to cumulative impacts on some vegetation communities, depending on the number and location of other developments in the region.	Dry wash, dry wash woodland, saguaro cactus, and ironwood (including those outside of washes) vegetation communities within the SEZ and associated new roads or transmission lines will be avoided to the extent practicable. A buffer area will be maintained around dry washes, dry wash woodland, to reduce the potential for impacts. Travel through weed-infested areas will be avoided; vehicles and equipment will be inspected and cleaned to avoid spreading weeds; ground disturbance will be limited, soil conditions that promote weed germination and establishment will be avoided, seed and plant parts will be disposed of. See programmatic design features at http://blmsolar.anl.gov/documents/docs/peis/programmatic-design-features/Ecological_Resources.pdf	Appropriate engineering controls will be used to minimize impacts on dry wash vegetation communities, including downstream occurrences, resulting from surface water runoff, erosion, sedimentation, altered hydrology, accidental spills, or fugitive dust deposition to these habitats. Appropriate buffers and engineering controls will be determined through agency consultation. Groundwater withdrawals will be limited to reduce the potential for indirect impacts on groundwater-dependent communities, such as, microphyll (paloverde/ ironwood) communities, or riparian habitats along the Gila or Hassayampa Rivers.	Yes. Development would result in direct removal or disturbance of these native plant communities, special soil environments, and the ecosystem services they provide.

Table A-3. (Cont.)

Resource/Issue	Gillespie Solar Energy Zone Impacts[1]	On-site Mitigation[2]		Residual Adverse Impacts[3]? (include justification)
		Avoidance	Minimization	
Ecology: Riparian Areas Section 8.3.10	**Direct:** Development will adversely affect characteristic vegetation (e.g., creosote bush, white bursage, cactus, paloverde, and ironwood) through destruction and loss of habitat. Development, including vegetation removal, land clearing, grading, changes in surface water flow, and dust deposition may alter soils and vegetation communities and result in the establishment of invasive species and noxious weeds within the SEZ. **Indirect:** Loss of native vegetation due to dust deposition from construction and operations, increased surface water runoff and related erosion, or through the introduction of invasive species. Establishment of noxious weeds in the SEZ may result in their spreading to adjacent areas. **Cumulative:** Solar energy development could be a contributor to cumulative impacts on some vegetation communities, depending on the number and location of other developments in the region.	Dry washes, playas, and wetlands within the SEZ and dry washes within the access road corridor will be avoided to the extent practicable. A buffer area will be maintained around wetlands, playas, and dry washes to reduce the potential for impacts. Appropriate engineering controls will be used to minimize impacts on dry wash, dry wash woodland and chenopod scrub, including downstream occurrence, resulting from surface water runoff, erosion, sedimentation, altered hydrology, accidental spills, or fugitive dust deposition to these habitats. Appropriate buffers and engineering controls will be determined through agency consultation. See programmatic design features at http://blmsolar.anl.gov/documents/docs/peis/programmatic-design-features/Ecological_Resources.pdf	See programmatic design features. Groundwater withdrawals will be limited to reduce the potential for dependent communities, such as, microphyll (paloverde/ ironwood) communities, or riparian habitats along the Gila or Hassayampa Rivers.	Maybe. Depends on the degree of avoidance and engineering controls. Development may alter ephemeral stream channels that can impact flooding and debris flows during storms, groundwater recharge, ecological habitats, and riparian vegetation communities. Reductions to the connectivity of these areas with existing surface waters and groundwater could limit water availability and thus alter the ability of the area to support vegetation and aquatic species. This could reduce overall stability of the natural landscape.

Table A-3. (Cont.)

Resource/Issue	Gillespie Solar Energy Zone Impacts[1]	On-site Mitigation[2]		Residual Adverse Impacts[3]? (include justification)
		Avoidance	Minimization	
Ecology: Invasive & Noxious Weeds Section 8.3.10	**Direct:** Development, including vegetation removal, land clearing, grading, changes in surface water flow, and dust deposition may alter soils and vegetation communities and result in the establishment of invasive species and noxious weeds within the SEZ. **Indirect:** Loss of native vegetation due to dust deposition from construction and operations, increased surface water runoff and related erosion, or through the introduction of invasive species. Establishment of noxious weeds in the SEZ may result in their spreading to adjacent areas. **Cumulative:** Solar energy development could be a contributor to cumulative impacts on some vegetation communities, depending on the number and location of other developments in the region.	Travel through weed infested areas will be avoided; vehicles and equipment will be inspected and cleaned to avoid spread of weeds; ground disturbance will be limited, creation of soil conditions that promote weed germination and establishment will be avoided, and disposal of seed and plant parts will be disposed of to reduce invasive species establishment. See programmatic design features at http://blmsolar.anl.gov/documents/docs/peis/programmatic-design-features/Ecological_Resources.pdf	Impacts will be minimized through development of a Weed Management Plan and use of weed-free seed to support re-vegetation efforts, control invasive species, and prevent increase in fires. See programmatic design features.	Maybe. On-site mitigation will reduce, but not eliminate, the potential for invasive species. The degree of disturbance creates a significant opportunity for the establishment of invasive species and weeds.
Ecology: Terrestrial Wildlife and Aquatic Biota Section 8.3.11	**Direct:** Loss of habitat and connectivity for several species of amphibians, reptiles, mammals, bats, and invertebrates. Remaining habitat in SEZ could be of reduced value for some species. Ground disturbance, fugitive dust generated by project activities, lighting, vegetation clearing, spread of invasive species, accidental spills, harassment, and ephemeral wash loss could impact wildlife within the SEZ. Impacts from noise on wildlife could occur, especially on bat species, if the SEZ is located near any bat roosts. **Indirect:** Outside the SEZ, impacts could occur from habitat loss or modification related to groundwater depletions, surface runoff, dust, noise, lighting, or accidental spills. **Cumulative:** Cumulative effects from all future development in the region on some species could be moderate, depending on the type, number, and location of other developments in the region. **Data Gaps:** Impacts on terrestrial wildlife from construction noise would have to be considered on a project-specific basis, especially for bat species.	Wetlands identified during site-specific fieldwork will be avoided to the extent possible. See programmatic design features at http://blmsolar.anl.gov/documents/docs/peis/programmatic-design-features/Ecological_Resources.pdf	The fencing around the solar energy development should not block the free movement of mammals, particularly big game species. Appropriate engineering controls should be implemented to minimize the amount of contaminants and sediment entering wetlands and washes within the SEZ. See programmatic design features.	Yes. Development of the Gillespie SEZ will likely impact up to 2,618 acres of wildlife habitat. Level of site grading and disturbance to native vegetation would be primary driver of residual impact for full build-out of SEZ.

Table A-3. (Cont.)

Resource/Issue	Gillespie Solar Energy Zone Impacts[1]	On-site Mitigation[2]		Residual Adverse Impacts[3]? (include justification)
		Avoidance	Minimization	
Ecology: Migratory Birds Section 8.3.11.2	**Direct:** Loss of individuals, habitat, and connectivity for several species protected under the Migratory Bird Treaty Act. Noise, lighting, and vegetation clearing could impact migratory birds using the SEZ. Water birds could be attracted to solar fields (because they look like water) and may collide with solar panels. Burning of wings in the solar radiation field between heliostats and power towers has been observed. There may also be impacts to night sky that may alter bird migratory behavior and habitat use. Priority migratory bird species that may occur on or near the SEZ include Gila woodpecker, gilded flicker, and LeConte's thrasher.[24] **Indirect:** Outside the SEZ, impacts could occur from habitat loss. **Cumulative:** Impacts to migratory birds could occur; depending on the type, number, and location of other developments in the region. **Data Gaps:** Additional research needed on solar development impacts on migratory birds. Impacts on migratory birds from construction noise would have to be considered on a project-specific basis.	Effects to individual migratory birds and bird nests can be avoided by not constructing during the breeding season. Timing limitation should be enforced from May 15—July 15 for any surface disturbing activities to protect migratory bird nesting and brood rearing. If construction takes place during the breeding season, nest surveys will be conducted. See programmatic design features at http://blmsolar.anl.gov/documents/docs/peis/programmatic-design-features/Ecological_Resources.pdf	Recommend implementation of technologies that minimize the amount of reflective surfaces, or alter how the surfaces are perceived by wildlife, that will reduce the "lake effect" in attracting migratory birds and other wildlife. See programmatic design features.	Yes. Development of the Gillespie SEZ will likely impact up to 2,618 acres of migratory bird habitat. Some level of bird injury/fatality has been observed for all types of solar facilities (through collisions with equipment or from burns). Research is ongoing to quantify impacts and identify effective mitigation measures.

[24] Priority migratory bird species for the SEZ were determined based on those species discussed in the Lower Sonoran RMP, the distribution of Arizona Natural Heritage Program tracked species, and USFWS Birds of Conservation Concern in the Arizona Habimap tool (http://www.habimap.org/).

Table A-3. (Cont.)

Resource/Issue	Gillespie Solar Energy Zone Impacts[1]	On-site Mitigation[2]		Residual Adverse Impacts[3]? (include justification)
		Avoidance	Minimization	
Ecology: Plant Special Status Species Section 8.3.12	**Direct:** No Endangered Species Act (ESA)-listed or BLM-listed plant species have been identified that have suitable habitat within the SEZ. Ground disturbance, land clearing and grading, and fugitive dust generated by project activities would result in loss of special status plant species habitat, if present, and might result in loss of individual plants. **Indirect:** Indirect impacts to individuals and habitat outside of the SEZ could occur from groundwater depletions, surface runoff, dust, noise, lighting, or accidental spills. Suitable habitat for two BLM-sensitive plant species has been identified on or near the SEZ. However, indirect impacts on these species would be small, with less than 1 percent of these species habitat in the SEZ region lost. Potential impacts from groundwater withdrawals. **Cumulative:** There would be no cumulative impacts on special status plant species unless they are discovered during pre-disturbance surveys (cumulative impacts then might be due to habitat destruction and overall development and fragmentation of the area). **Data Gaps:** Pre-disturbance surveys are required to identify the presence and abundance of special status species.	Based on data from pre-disturbance surveys, disturbance to occupied habitats would be avoided to the extent practicable. Desert playa, wash habitats, sand dunes, transport systems, woodlands, rocky cliffs, and outcrops will be avoided to the extent practicable. See programmatic design features at http://blmsolar.anl.gov/documents/docs/peis/programmatic-design-features/Ecological_Resources.pdf	If avoidance is not possible for some species, translocation of individuals from areas of direct effects or compensatory mitigation may be employed. See programmatic design features.	No, unless special status plant species are discovered during pre-disturbance surveys. There are no known SSS plant species within the SEZ.

Table A-3. (Cont.)

Resource/Issue	Gillespie Solar Energy Zone Impacts[1]	On-site Mitigation[2]		Residual Adverse Impacts[3]? (include justification)
		Avoidance	Minimization	
	Direct: Ground disturbance, land clearing and grading, and fugitive dust generated by project activities would result in loss of special status animal species habitat, if present, and might result in loss of individual animals. Impacts from noise on special status wildlife could also occur. Solar PEIS analyses indicated that development on the SEZ could directly disturb individuals or habitat for two candidate species for listing under the ESA (i.e., the **Sonoran desert tortoise**[25] and the Tucson shovel-nosed snake), and six BLM sensitive special status animal species (lowland leopard frog, Mexican rosy boa, Sonoran bald eagle, snowy egret, Western burrowing owl, and California leaf-nosed bat). Subsequent BLM analyses show that Gillespie SEZ is actually outside the occurrence area of the following species: Tucson shovel-nosed snake, lowland leopard frog, Sonoran bald eagle, and snowy egret. Impacts to these species will not be further considered.		If avoidance is not possible for some species, translocation of individuals from areas of direct effects or compensatory mitigation may be employed.	
Ecology: Animal Special Status Species Section 8.3.12	No Category 1, 2, or 3 desert tortoise habitat has been identified by BLM within the SEZ; however, Category 2 desert tortoise habitat occurs outside the SEZ adjacent to the southern border. Although no Categorized desert tortoise habitat occurs on the SEZ, desert tortoises may still occur in lower quality habitat on the SEZ where they may be directly impacted by solar development.	Compliance with the Bald & Golden Eagle Protection Act would be ensured and Eagle Take Guidance would be followed (if necessary). Based on data from pre-disturbance surveys, disturbance to suitable habitats would be avoided to the extent practicable. See programmatic design features at http://blmsolar.anl.gov/documents/docs/peis/programmatic-design-features/Ecological_Resources.pdf	Regarding avoidance and minimization, consultation with the USFWS will be conducted to address the potential for impacts on ESA-listed and proposed species and to identify mitigation measures for implementation. Groundwater withdrawals will be avoided or minimized to reduce or eliminate impacts on nine special status species. See programmatic design features.	Yes. Animal SSS along with other wildlife represent a basic component of the ecosystem. Level of site grading and disturbance to native vegetation would be the primary driver of residual impact to functional habitat for full build out of the SEZ.
	Indirect: Indirect impacts to individuals and animal habitat outside of the SEZ could occur due to groundwater depletions, surface runoff, dust, noise, lighting, or accidental spills. Suitable habitat for 3 ESA-listed (**southwestern willow flycatcher, western yellow-billed cuckoo,** and **Yuma clapper rail**) and 4 BLM-sensitive animal species (roundtail chub, ferruginous hawk, great egret, and Western red bat) occurs near the SEZ. For groundwater dependent species, impacts could range from small to large depending on groundwater use for development.			
	Cumulative: There could be cumulative impacts on some special status animal species due to habitat destruction and overall development and fragmentation of the area.			
	Data Gaps: Pre-disturbance surveys are required to identify the presence and abundance of special status species.			

---

[25] Species in bold text have been recorded within 5 miles (8 km) of the SEZ.

Regional Mitigation Strategy for the Arizona SEZs

Table A-3. (Cont.)

| Resource/Issue | Gillespie Solar Energy Zone Impacts[1] | On-site Mitigation[2] | | Residual Adverse Impacts[3]? (include justification) |
		Avoidance	Minimization	
Environmental Justice Section 8.3.20	**Direct:** There is a minority population within a 50-mile (80 km) radius of the SEZ, so any adverse impacts of solar projects could affect this population. There are no low-income populations within a 50-mile radius of the SEZ. Positive impacts are possible if solar facility-related employment increases. **Indirect:** None identified. **Cumulative:** Contributions from solar development in the SEZ would likely be small and would not be expected to significantly contribute to cumulative impacts on minority populations within the 50-mile geographic extent of effects.	See programmatic design features at http://blmsolar.anl.gov/documents/docs/peis/programmatic-design-features/Environmental_Justice.pdf	See programmatic design features.	No.
Hydrology: Surface Water Section 8.3.9	**Direct:** Land clearing, land leveling, vegetation removal, and spills and runoff associated with development of the SEZ could increase surface runoff, reduce infiltration/recharge, cause loss of ephemeral stream networks, cause a reduction in evapotranspiration rates, increase sediment transport (by water), change sediment transport (by wind), and degrade water quality. No perennial surface water features have been identified within the SEZ. The SEZ is located on sloping land containing more than 29 miles (46 km) of intermittent/ephemeral wash tributaries to Centennial Wash (a tributary to the Gila River). Based on an evaluation of data in the Sonoran Desert REA, ephemeral drainages with high potential for water erosion occur on the SEZ. **Indirect:** Indirect impacts from development and groundwater use on ephemeral and perennial surface water features could occur. **Cumulative:** Alterations to ephemeral stream networks can alter groundwater recharge and surface runoff processes potentially impacting the basin-scale water balance and water quality aspects of water features receiving surface runoff. **Data Gaps:** Project siting and design will need to consider impacts to the washes located in the SEZ.	See programmatic design features at http://blmsolar.anl.gov/documents/docs/peis/programmatic-design-features/Water.pdf	See programmatic design features.	Yes. Hydrology is a basic component of the ecosystem. Reconfiguration of topography for solar development would have residual impacts to surface hydrology with potential impacts on other resources, including vegetation.

Table A-3. (Cont.)

Resource/Issue	Gillespie Solar Energy Zone Impacts[1]	On-site Mitigation[2]		Residual Adverse Impacts[3]? (include justification)
		Avoidance	Minimization	
Hydrology: Water Quality and Groundwater Availability Section 8.3.9	**Direct:** If water intensive technology is used, groundwater withdrawals for development may cause declines in groundwater elevations that can impact water availability for surface water features, vegetation, ecological habitats, regional groundwater flow paths, and other groundwater users in the basin. A riverine wetland is located just inside the southeast corner of the SEZ. The Gillespie SEZ is in the Lower Hassayampa groundwater basin, where the primary aquifer is composed of basin-fill alluvium deposits. **Indirect:** Possible groundwater withdrawals for solar energy facilities have the potential to affect other groundwater users in the basin. **Cumulative:** Cumulative impacts on groundwater could occur depending on the type, number, and location of other developments in the region.	Groundwater analyses suggest that full build-out of wet-cooled technologies is not feasible. See programmatic design features at http://blmsolar.anl.gov/documents/docs/peis/programmatic-design-features/Water.pdf	The SEZ is located in Water Protection Zone 3 and new water uses and withdrawals are restricted to panel washing and sanitary uses only.[26] For mixed-technology development scenarios, any proposed wet-cooled projects would be required to retire existing groundwater uses and utilize water conservation practices. See programmatic design features.	No. It is possible for impacts on groundwater aquifers to be avoided or minimized.

[26] Unavoidable adverse impacts are possible if groundwater is used. However, wet cooling was not considered a feasible option in the Solar PEIS ROD. Additional restrictions identified in the RDEP ROD for Water Protection Zone 3 would further limit the potential for residual impacts to occur.

Regional Mitigation Strategy for the Arizona SEZs

Table A-3. (Cont.)

Resource/Issue	Gillespie Solar Energy Zone Impacts[1]	On-site Mitigation[2]		Residual Adverse Impacts[3]? (include justification)
		Avoidance	Minimization	
Lands & Realty Section 8.3.2	**Direct:** Full development of Gillespie SEZ would disturb 2,618 acres (11 km^2). A Right-of-Way (ROW) for the existing Agua Caliente Road (29 acres [0.1 km^2] of the SEZ) would be protected as a requirement of any solar development proposal. The road cuts the SEZ area into smaller portions and provides public access through the site. To avoid these issues, relocation of the road may be considered as part of a site development plan and would require additional analysis. **Indirect:** Impacts due to altering uses on public, state, and private lands in the vicinity of the SEZ. Examples include increased traffic and increased access to previously remote areas also could change the overall character of the landscape. **Cumulative:** Cumulative effects on land use could occur through impacts on land access and use for other purposes particularly if additional solar development occurred in the region. However, projects within the SEZ would make only a small contribution to cumulative impacts because of its relatively small size.	See programmatic design features at http://blmsolar.anl.gov/documents/docs/peis/programmatic-design-features/Lands_and_Realty.pdf	Priority consideration should be given to using the existing Agua Caliente Road to provide construction and operations access to the SEZ. Any potential impacts on the existing county road should be discussed with the county. See programmatic design features.	No. By regulation, any new activity must occur in deference to existing rights. Thus, potential impacts have been avoided.
Livestock Grazing Section 8.3.4.1	**Direct:** The SEZ includes small portions of four grazing allotments. The percentage of three of the four allotments that intersect the SEZ is less than 1.5 percent of each allotment. Impacts on the three allotments would be small. Potential impacts on the fourth ephemeral allotment could not be determined at the time of the Final Solar PEIS. **Indirect:** None identified. **Cumulative:** Other development in the area of the SEZ could result in cumulative impacts on grazing. However, the contribution of such effects from projects within the SEZ would be minimal due to the small area affected.	See programmatic design features at http://blmsolar.anl.gov/documents/docs/peis/programmatic-design-features/Rangeland_Resources.pdf	See programmatic design features.	Maybe. Residual impacts to be evaluated based on locations of development within the SEZ and project-level NEPA.

A-51

Table A-3. (Cont.)

Resource/Issue	Gillespie Solar Energy Zone Impacts[1]	On-site Mitigation[2]		Residual Adverse Impacts[3]? (include justification)
		Avoidance	Minimization	
Military & Civilian Aviation Section 8.3.6	**Direct:** There is one military training route (MTR) above the SEZ; the MTR has a 300-foot (91-m) above-ground-level operating limit. The military says that the construction of solar or related facilities in excess of 250 feet (76 m) tall could interfere with military training activities and be a safety concern. Buckeye and Gila Bend Municipal Airports are 15 miles (42 km) northeast and 20 miles (32 km) south-southeast, respectively. Neither has regularly scheduled passenger or freight service. **Indirect:** None identified. **Cumulative:** Solar development occurring throughout the region, which is largely undeveloped, could result in small cumulative effects on the system of MTRs. Such effects would be limited by mitigations developed in consultation with the military.	See programmatic design features at http://blmsolar.anl.gov/documents/docs/peis/programmatic-design-features/Military_Civilian_Aviation.pdf	Coordination with Federal Aviation Administration and the military will be required on a project-specific basis to ensure that solar facilities do not interfere with operations. See programmatic design features.	Maybe (with respect to MTRs). Residual impacts will be evaluated based on coordination with the military and project-level NEPA.
Minerals Section 8.3.8 and Section 8.3.24 of the Final Solar PEIS	**Direct:** There is one placer mining claim in the very northwestern portion of the SEZ, about 260 acres (1 km²) in size. No solar development would be possible within this area without the claimant's agreement or unless the claim is ruled to be invalid. The SEZ has been withdrawn from mineral entry for a period of 20 years, precluding impacts from many types of mining activities. **Indirect:** None identified. **Cumulative:** The specific locations of mining claims will be identified during project-specific analyses.	See programmatic design features at http://blmsolar.anl.gov/documents/docs/peis/programmatic-design-features/Mineral_Resources.pdf	See programmatic design features.	No. The existing mining claim is a prior existing right and, if valid, likely would preclude development of the portion of the SEZ in which the claim is located.

Table A-3. (Cont.)

Resource/Issue	Gillespie Solar Energy Zone Impacts[1]	On-site Mitigation[2]		Residual Adverse Impacts[3]? (include justification)
		Avoidance	**Minimization**	
	Direct: Tribes are likely to have major concerns about the impact of development on water resources and on traditional plants and animal resources. Removal of cultural resources is a concern to the tribes. Development of solar could impact access to a sacred area or place of traditional cultural importance.	Known human burial sites and rock art (panels of petroglyphs and/or pictographs) will be avoided. The BLM will consult with Indian tribes regarding the potential for unanticipated human remains and associated cultural items (as defined under the Native American Graves Protection and Repatriation Act), before a solar project is authorized. The purpose will be to discuss general guidance on treatment of cultural items.		
Native American Concerns Section 8.3.18	**Indirect:** General habitat loss with vegetation clearing and water reduction that could affect species and ecosystem health.	Springs and other water sources that are or may be sacred or culturally important, culturally important plant and wildlife species, and visual intrusion on sacred sites will be avoided to the extent possible.	See programmatic design features.	Yes. Consultation on project applications will determine whether regional mitigation for Native American Concerns is warranted.
	Cumulative: Development of solar energy facilities in combination with the development of other planned and foreseeable projects in the area would likely reduce the traditionally important plant and animal resources available to the tribes. Although some of these plant species are abundant, any level of impact may be of concern for the tribes.	EO13007 requires executive branch agencies to accommodate access to and ceremonial use of Indian sacred sites and to avoid adverse impacts to the physical integrity of such a site. Because solar facilities will be fenced and security procedures will limit or eliminate access, if a sacred site was declared, it may not be possible to mitigate impacts.	Mitigate onsite to comply with EO13007. BLM could facilitate the harvest of creosote prior to ground disturbance for instance, if identified as a concern.	
	Data Gaps: Documentation of an archaeological survey of the entire SEZ is currently being completed and results will be shared with the tribes. Government-to-Government consultation for projects will be required to determine issues of Native American concern.	See programmatic design features at http://blmsolar.anl.gov/documents/docs/peis/programmatic-design-features/Native_American_Concerns.pdf		

Table A-3. (Cont.)

Resource/Issue	Gillespie Solar Energy Zone Impacts[1]	On-site Mitigation[2]		Residual Adverse Impacts[3]? (include justification)
		Avoidance	Minimization	
Paleontology Section 8.3.16	**Direct:** The SEZ is in an area classified as Potential Fossil Yield Classification (PFYC) Class 3b. It has a low to undetermined potential for paleontological resources. **Indirect:** None identified. **Cumulative:** Cumulative impacts would be dependent on whether significant resources are found within the SEZ and in additional project areas in the region. **Data Gaps:** Potential for impacts is unknown. A more detailed assessment of the geological deposits of the SEZ is needed to determine whether a paleontological survey is warranted for a specific project.	See programmatic design features at http://blmsolar.anl.gov/documents/docs/peis/programmatic-design-features/Paleo.pdf	The BLM will be notified immediately upon discovery of fossils. Work will be halted at the fossil site and continued elsewhere until qualified personnel, such as a paleontologist, can visit the site. He/she will determine if the site is significant and make recommendations for collection or other resource protection, if warranted. See programmatic design features.	No. Design features will reduce the risk that any paleontological resources that are discovered will be destroyed.
Public Access and Recreation Section 8.3.5	**Direct:** Development may impact recreational activities that occur within the SEZ boundary and surrounding specially designated areas or the Saddle Mountain Special Recreation Management Area (SRMA). Public access to areas south of the SEZ could be adversely impacted. It is anticipated that some users of portions of the nearby wilderness areas (e.g., Gila Bend Wilderness Area) may choose to move their activities farther away from solar energy facilities. **Indirect:** Indirect effects would occur primarily on lands near the solar facilities and would result from the change in the overall character of undeveloped BLM-administered lands to an industrialized, developed area. People seeking more rural or primitive surroundings for recreation may go elsewhere. **Cumulative:** Multiple developments could reduce recreational opportunities in the vicinity of the SEZ.	Because of the potential for solar development to sever current access routes from the county road within the SEZ, legal access to the areas to the south should be maintained consistent with existing land use plans. See programmatic design features at http://blmsolar.anl.gov/documents/docs/peis/programmatic-design-features/Public_Access_and_Recreation.pdf	See programmatic design features.	Maybe.

Table A-3. (Cont.)

| Resource/Issue | Gillespie Solar Energy Zone Impacts[1] | On-site Mitigation[2] | | Residual Adverse Impacts[3]? (include justification) |
		Avoidance	Minimization	
Socio-economics Section 8.3.19	**Direct:** Impacts on the local economy as a result of expenditures of wages and salaries and the collection of state sales and income taxes. From 92 to 1,218 direct construction jobs and 5 to 91 direct operations jobs could be created (least for PV; most for parabolic trough facilities). Adverse impacts could occur due to the need for services required for project construction and operation (e.g., police, firefighters). **Indirect:** From 196 to 2,600 indirect construction jobs and 1 to 59 indirect operations jobs could be created. Impacts from project wages and salaries, and tax revenues subsequently circulating through the economy would be minor. **Cumulative:** Impacts overall are expected to be positive, through the creation of additional jobs and income. The negative impacts, including some short-term disruption of rural community quality of life, are expected to be small.	See programmatic design features at http://blmsolar.anl.gov/documents/docs/peis/programmatic-design-features/Socioeconomics.pdf	See programmatic design features. Additionally, on-site mitigation could include requiring developers to secure agreements for local government services as a condition of "Notice to Proceed".	No. Generally positive impacts expected.

Table A-3. (Cont.)

Resource/Issue	Gillespie Solar Energy Zone Impacts[1]	On-site Mitigation[2] Avoidance	On-site Mitigation[2] Minimization	Residual Adverse Impacts[3]? (include justification)
Soils/Erosion Section 8.3.7	**Direct:** Impacts on soil resources would occur mainly as a result of ground-disturbing activities (e.g., grading, excavating, and drilling), especially during construction. These include topsoil removal, soil compaction, soil horizon mixing, soil erosion and deposition by wind, soil erosion by water and surface runoff, sedimentation, and soil contamination. Soils within the SEZ are extremely gravelly sandy loams and very gravelly sandy loams typical of alluvial fan (and fan terrace) settings, likely to be impacted through compaction and erosion. Soil contamination from spills could occur. Based on an evaluation of data in the Sonoran Desert REA, the majority of the soils on the SEZ have high potential for wind erosion. Therefore, increased wind erosion is likely if grading is needed. **Indirect:** Disturbance of soil can lead to introduction of invasive species. **Cumulative:** Cumulative impacts would occur from the disturbance of several renewable energy projects, connecting linear facilities, and other projects in the vicinity of the SEZ, but would be limited through application of design features.	See programmatic design features at http://blmsolar.anl.gov/documents/docs/peis/programmatic-design-features/Soil_Geologic_Hazards.pdf	Construction crews should be educated to stay on designated roads and minimize the construction of new roads to minimize soil disturbance and compaction. See programmatic design features.	Yes. Soils represent a basic component of the ecosystem. Solar development on the SEZ is expected to result in a residual loss of sensitive soils and soil functions.

Table A-3. (Cont.)

Resource/Issue	Gillespie Solar Energy Zone Impacts[1]	On-site Mitigation[2]		Residual Adverse Impacts[3]? (include justification)
		Avoidance	Minimization	
Specially Designated Areas and Lands with Wilderness Characteristics Section 8.3.3	**Direct:** Specially Designated Areas (SDAs) within 25 miles (40 km) of the SEZ could be visually impacted by solar development. Moderate to strong visual contrasts could be experienced in the Signal Mountains and Woolsey Peak Wilderness Areas (WAs), the closest of which is approximately 2 mi (3 km) from the SEZ. Impacts could include adverse visual effects on the viewshed (including impacts on the night sky viewing) and fragmentation of biologically linked areas. Minimal visual impacts are anticipated at Sonoran Desert National Monument, Juan Bautista de Anza National Historic Trail, Big Horn Mountains, Eagletail Mountains, Hummingbird Springs, North Maricopa Mountains, and South Maricopa Mountains WAs. There are no undesignated areas with wilderness characteristics near the SEZ. **Indirect:** None identified. **Cumulative:** Increased development and visual clutter in general in the surrounding areas, reduced local and regional visibility due to construction-related air particulates, light pollution, road traffic, and impacts on wildlife and plants may result in cumulative effects on SDAs.	See programmatic design features at http://blmsolar.anl.gov/documents/docs/peis/programmatic-design-features/SDAs_and_LWC.pdf	See programmatic design features.	Yes. Residual impacts will be evaluated based on locations of development within the SEZ and project level NEPA.
Transportation Section 8.3.21	**Direct:** Development will add traffic to existing roads serving the area. The volume of traffic on Old U.S. 80 could represent an increase in traffic of about 200 percent during construction. Local roads would also be impacted. **Indirect:** None identified. **Cumulative:** Cumulative impacts to traffic could occur with multiple developments in the region.	See programmatic design features at http://blmsolar.anl.gov/documents/docs/peis/programmatic-design-features/Transportation.pdf	Local roads such as Old U.S. 80 would require improvements to accommodate additional traffic. See programmatic design features.	No. Through a combination of avoidance, design features, and the establishment of alternative access routes to these areas, the potential impacts can be adequately mitigated onsite.

Table A-3. (Cont.)

Resource/Issue	Gillespie Solar Energy Zone Impacts[1]	On-site Mitigation[2]		Residual Adverse Impacts[3]? (include justification)
		Avoidance	Minimization	
Visual Section 8.3.14	**Direct:** The Visual Resource Inventory (VRI) values for the SEZ and immediate surroundings are VRI Class III, indicating moderate visual values. Development will adversely impact visual resources and may impact night skies. However, the Visual Resource Management (VRM) class is IV and allows for development. The Solar PEIS identified moderate to strong visual contrasts for some viewpoints within the Signal Mountain WA, Woolsey Peak WA, and the Saddle Mountain SRMA, as well as within the community of Arlington. Westbound travelers on Agua Caliente Road, a BLM-proposed backcountry byway and a scenic, high-use travel corridor would be subject to large to very large visual contrasts from solar facilities within the SEZ as they approached Agua Caliente Road from Old U.S. 80. **Indirect:** None identified. **Cumulative:** If several projects become visible from one location, or in succession as viewers move through the landscape (such as driving on local roads), these cumulative impacts may make the area less visually appealing.	See programmatic design features at http://blmsolar.anl.gov/documents/docs/peis/programmatic-design-features/Visual.pdf	Beyond those required for basic facility and company identification for safety, navigation, and delivery purposes, commercial symbols or signs and associated lighting on buildings and other structures should be prohibited. See programmatic design features.	Yes. While on-site mitigation would reduce visual contrasts caused by solar facilities within the SEZ, it would not likely reduce impacts to less than moderate or strong levels for nearby viewers.
Wild Horses and Burros Section 8.3.4.2	Because the Gillespie SEZ is 47 miles (76 km) or more from any wild horse and burro Herd Management Areas managed by the BLM and more than 50 miles (80 km) from any wild horse and burro territory administered by the U.S. Forest Service, solar energy development within the SEZ would not directly or indirectly affect wild horses and burros that are managed by these agencies.	Not applicable	Not applicable	No. The SEZ is not part of a herd management area, and no agency-managed horses or burros are known to exist in the area.

APPENDIX B

Conceptual Models

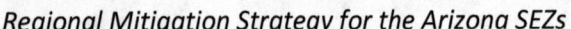

This page intentionally left blank

Tier 1 Conceptual Model
Sonoran Desert Ecoregion Model

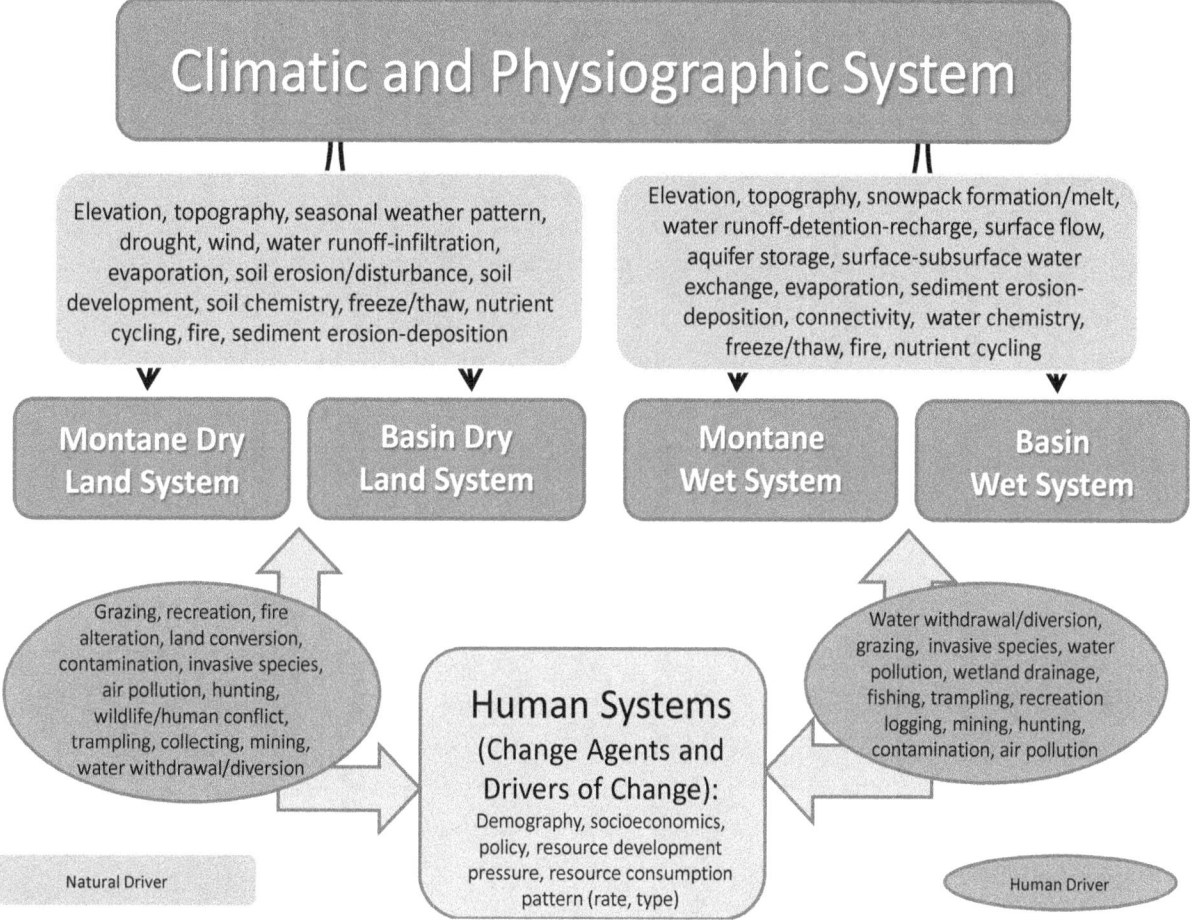

Regional Mitigation Strategy for the Arizona SEZs

Tier 2 Conceptual Model
Resource-Based Model

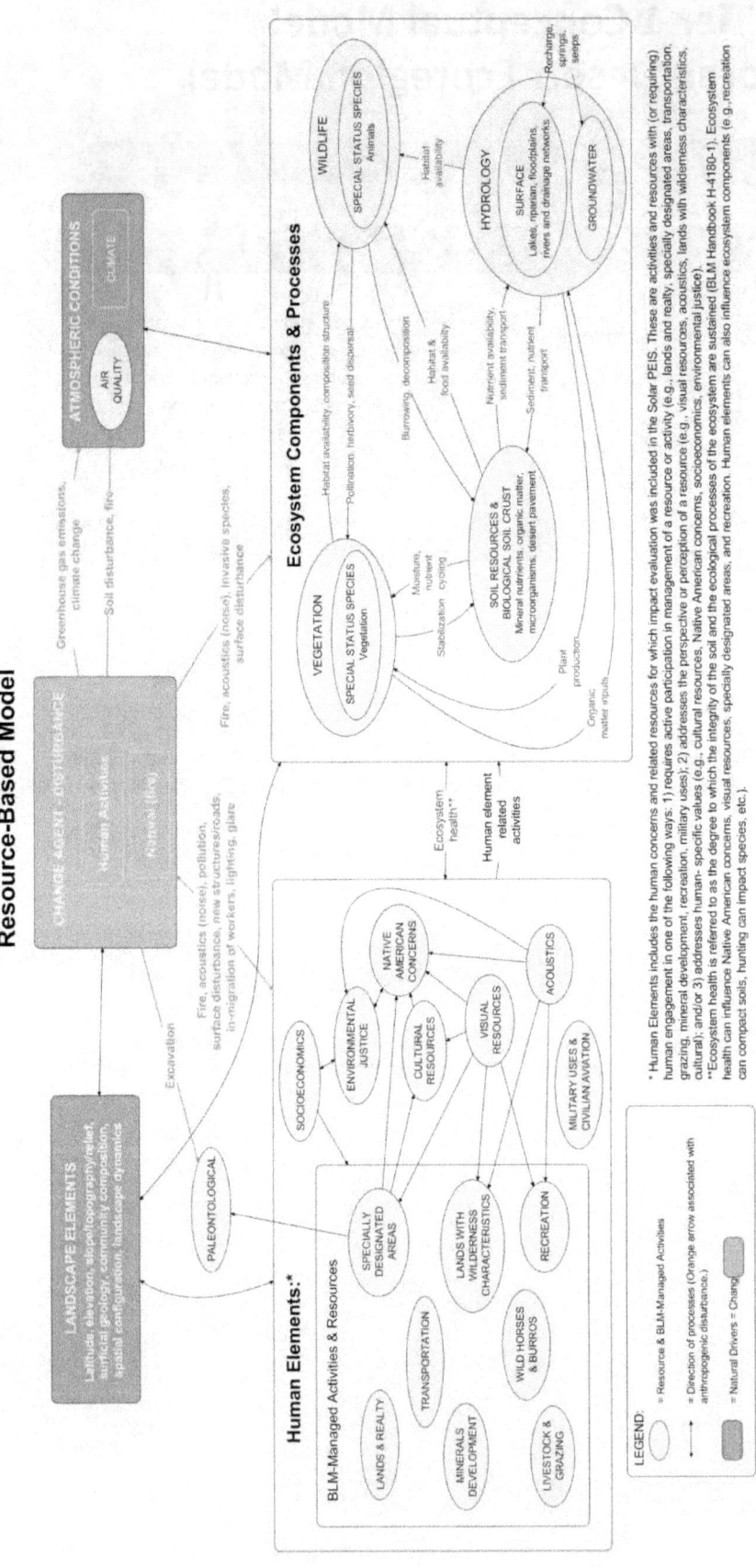

* Human Elements includes the human concerns and related resources for which impact evaluation was included in the Solar PEIS. These are activities and resources with (or requiring) human engagement in one of the following ways: 1) requires active participation in management of a resource or activity (e.g., lands and realty, specially designated areas, transportation, grazing, mineral development, recreation, military uses); 2) addresses the perspective or perception of a resource (e.g., visual resources, acoustics, lands with wilderness characteristics, cultural); and/or 3) addresses human- specific values (e.g., cultural resources, Native American concerns, socioeconomics; environmental justice).
** Ecosystem health is referred to as the degree to which the integrity of the soil and the ecological processes of the ecosystem are sustained (BLM Handbook H-4180-1). Ecosystem health can influence Native American concerns, visual resources, specially designated areas, and recreation. Human elements can also influence ecosystem components (e.g., recreation can compact soils, hunting can impact species, etc.).

Regional Mitigation Strategy for the Arizona SEZs

Tier 3 Conceptual Model
Agua Caliente SEZ Solar Development Model

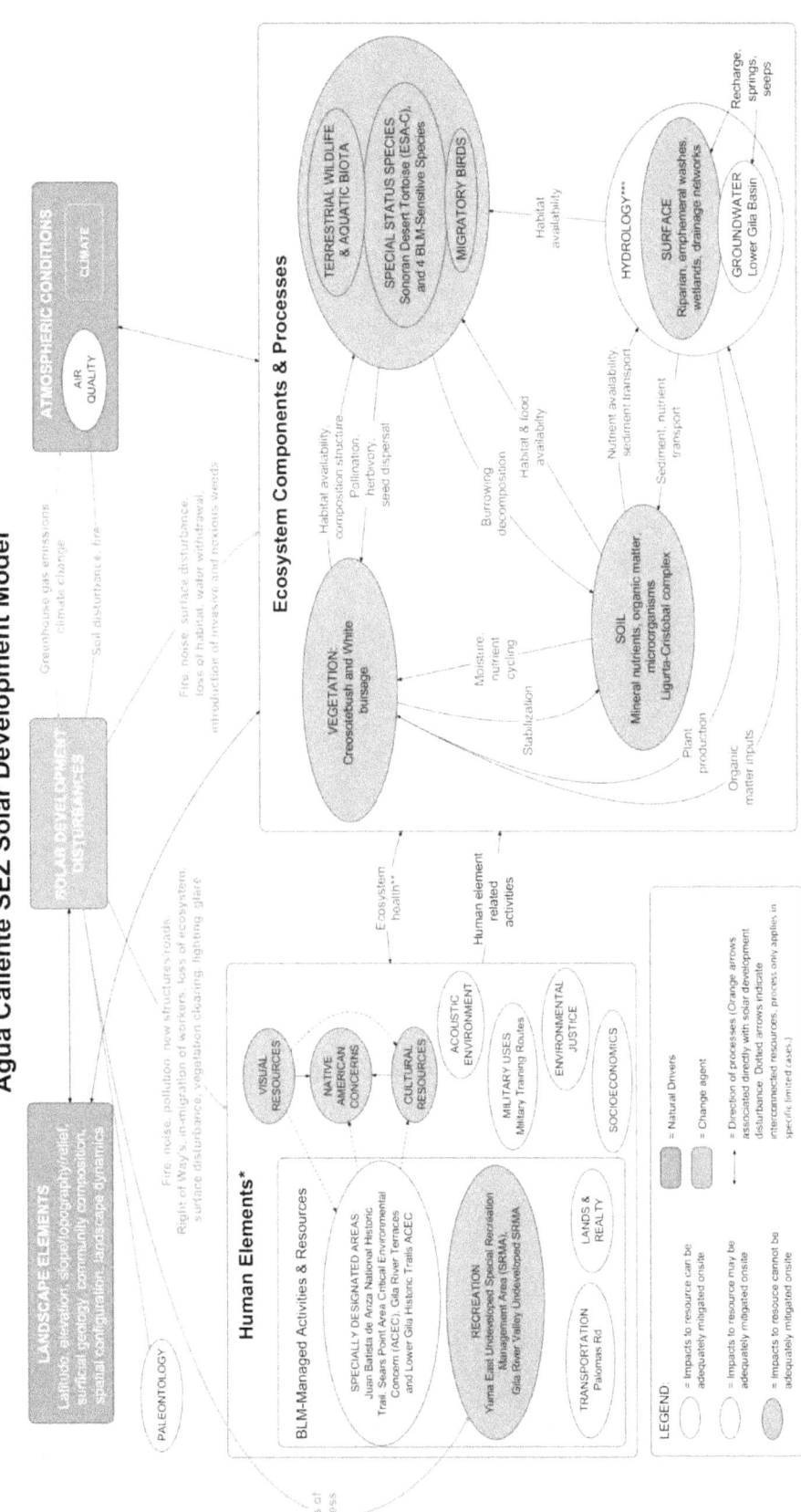

Regional Mitigation Strategy for the Arizona SEZs

Tier 3 Conceptual Model
Brenda SEZ Solar Development Model

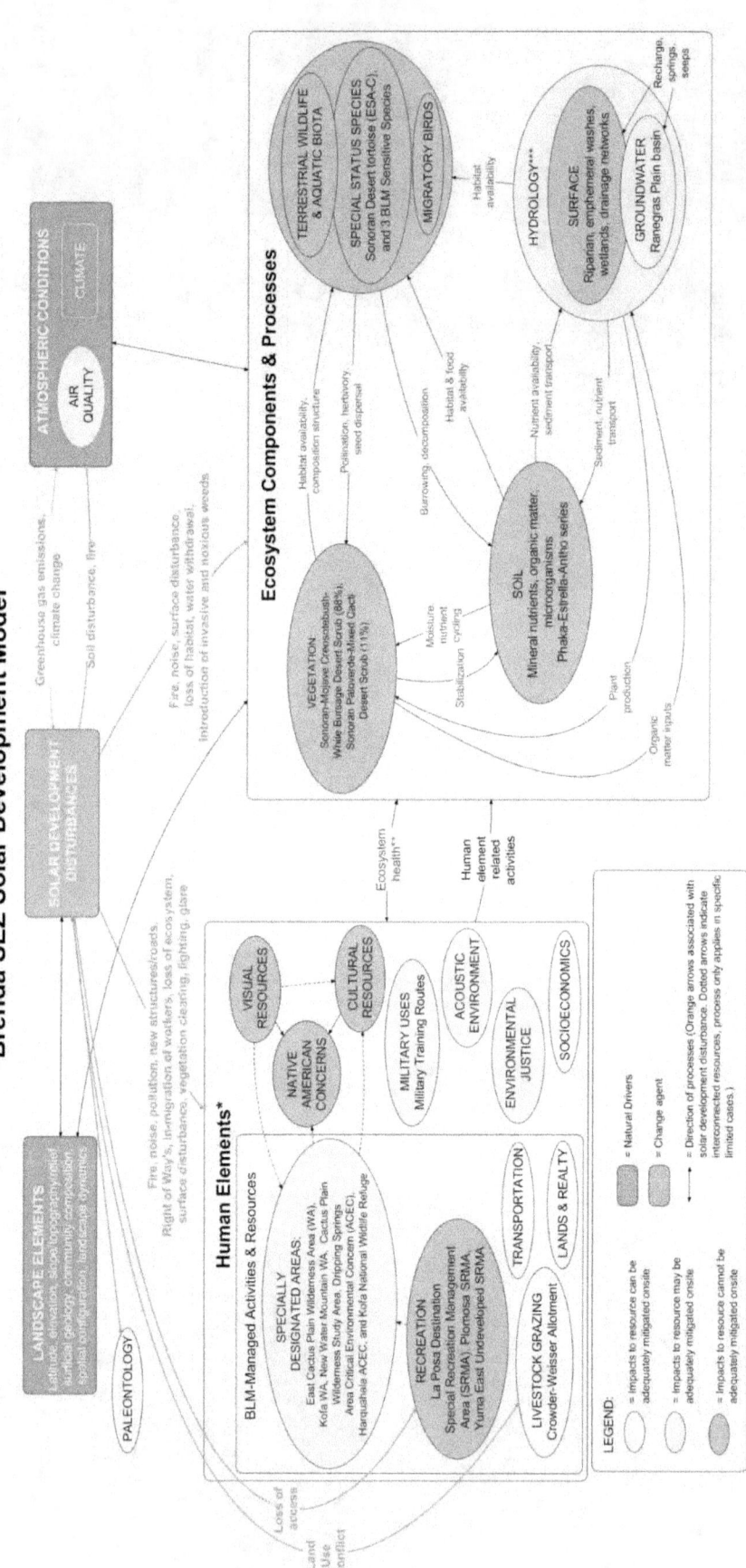

Regional Mitigation Strategy for the Arizona SEZs

Tier 3 Conceptual Model
Gillespie SEZ Solar Development Model

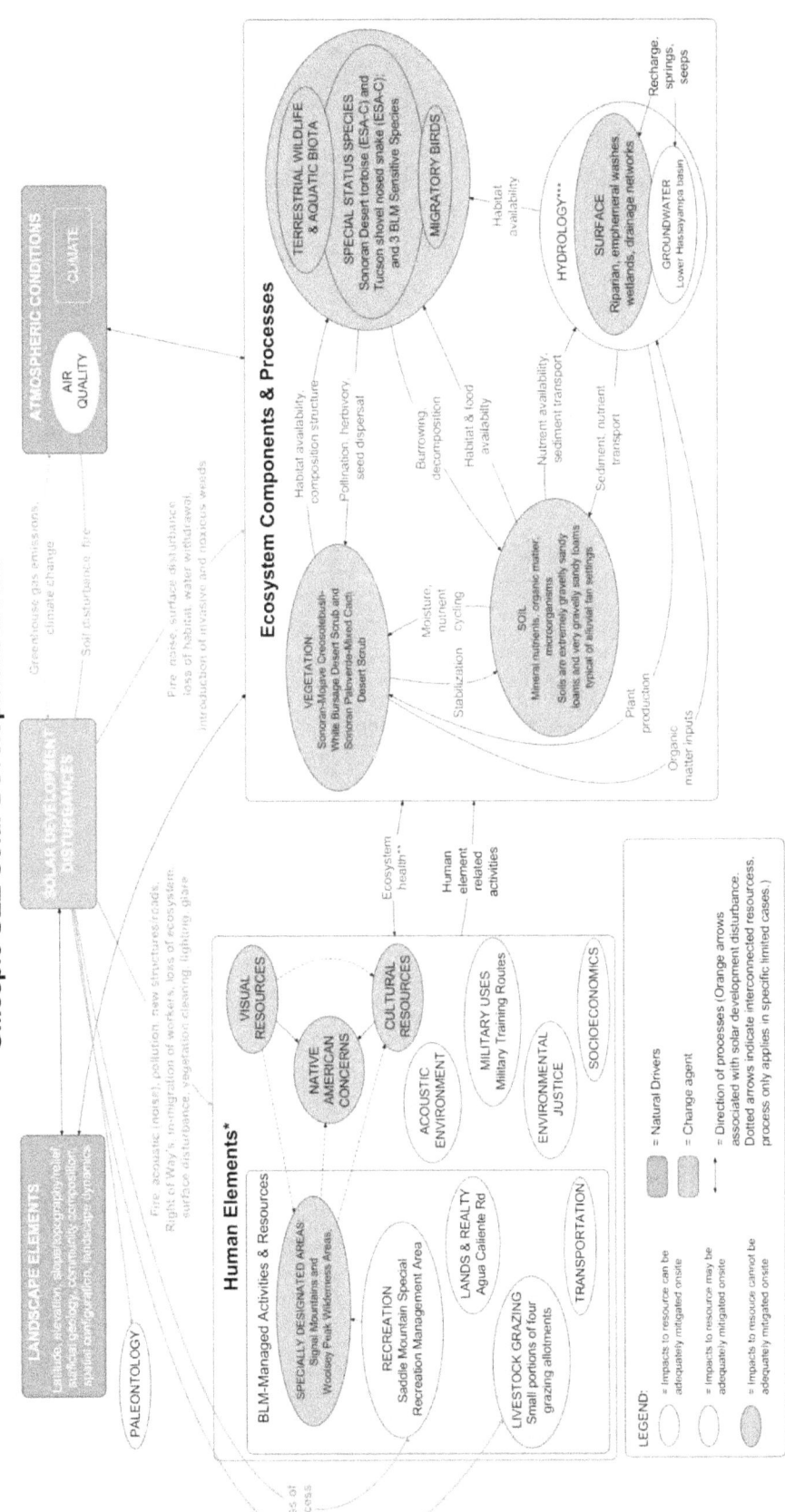

* Human Elements includes the human concerns and related resources for which impact evaluation was included in the Solar PEIS. These are activities and resources with (or requiring) human engagement in one of the following ways: 1) requires active participation in management of a resource or activity (e.g. lands and realty, specially designated areas, transportation, grazing, mineral development, recreation, military uses); 2) addresses the perspective or perception of a resource (e.g. visual resources, acoustics, lands with wilderness characteristics, cultural) and/or 3) addresses human- specific values (e.g. cultural resources, socioeconomics, environmental justice).
** Ecosystem health is referred to as the degree to which the integrity of the soil and the ecological processes of the ecosystem are sustained (BLM Handbook H-4180-1). Ecosystem health can influence Native American concerns visual resources, specially designated areas, and recreation. Human elements can also influence ecosystem components (e.g. recreation can compact soils, hunting can impact species, etc.)
*** Hydrologic impacts may occur due to changes in drainage and recharge patterns. these impacts can be mitigated onsite. Potential impacts to water availability will be mitigated onsite through the implementation of a net neutral use policy (water rights must be purchased).

This page intentionally left blank

APPENDIX C

**Summary Tables: Impacts that May Warrant Regional Mitigation
for the Three Arizona Solar Energy Zones
Agua Caliente, Brenda and Gillespie**

This page intentionally left blank

Regional Mitigation Strategy for the Arizona SEZs

Table C-1. Agua Caliente Solar Energy Zone – Summary Table: Impacts That May Warrant Regional Compensatory Mitigation

Agua Caliente Solar Energy Zone Resource/ Issue	Residual or Unavoidable Impact?	How certain is it that the residual impacts will occur?	How significant are the residual impacts onsite?	How significant are the residual impacts of developing the Agua Caliente SEZ in the region (Sonoran Desert)?	Role in the ecosystem?	Other Considerations	Are potential residual impacts likely to warrant regional mitigation?
Acoustics	Maybe.	Possible. Depends on technology and engineering controls.	Moderate. Construction-phase noise limited in duration; operation phase noise levels will be a permanent (30+ years) change.		Human Element		No. Generally impacts from solar development are expected to be temporary, localized, and readily mitigated.
Air Quality	Maybe (if site is graded).	Possible. Depends on whether the entire developable area (i.e., ~2,040 acres, 80% of the SEZ area) is graded.			Human Element	Ambient PM Levels.	No. Impacts are not expected to result in noncompliance with National Air Quality Standards.
Cultural	Yes.	Certain.	Moderate. Results of archaeological survey of the entire SEZ and some areas outside of the SEZ identified significant (eligible) sites, but most will be avoided by creation of non-development area within the SEZ.	Moderate. Results of archaeological survey of the entire SEZ and some areas outside of the SEZ identified significant (eligible) sites, but most will be avoided by creation of non-development area within the SEZ.	Human Element	Avoidance preferred for significant resources. Adequate mitigation would be determined during consultation and is dependent on the resources and their relative significance in the region.	Maybe. Impacts warranting mitigation to be evaluated in consultation with AZ State Historic Preservation Office (SHPO) and tribes.

C-3

Table C-1. (Cont.)

Agua Caliente Solar Energy Zone Resource/Issue	Residual or Unavoidable Impact?	How certain is it that the residual impacts will occur?	How significant are the residual impacts onsite?	How significant are the residual impacts of developing the Agua Caliente SEZ in the region (Sonoran Desert)?	Role in the ecosystem?	Other Considerations	Are potential residual impacts likely to warrant regional mitigation?
Ecology: Vegetation	Yes.	Certain.	Very. Expect the loss of all vegetation over the developable area of the SEZ, though mitigation may result in some remaining or replanted vegetation.	Decreasing trend in ecoregion.	Basic Component	Natural regeneration of native vegetation is slow in the Sonoran Desert.	Yes. Critical component of a functioning ecosystem.
Ecology: Riparian Areas	Maybe.	Possible. Depends on the degree of avoidance first then on engineering controls to address hydrologic impacts.		Decreasing trend in region, disproportionate historic impact to riparian areas.	Basic Component	Hydrologic impacts will affect riparian vegetation thus affect migratory birds, habitat for other wildlife—e.g., thermal cover for deer.	Yes.
Ecology: Invasive & Noxious Weeds	Maybe.	Possible. Depends on degree of vegetation disturbance and adequacy of Design Features.	Low.			Impacts will be minimized through development of a Weed Management Plan and use of weed-free seed to support re-vegetation efforts, control invasive species, and prevent increase in fires.	No.

Table C-1. (Cont.)

Agua Caliente Solar Energy Zone Resource/ Issue	Residual or Unavoidable Impact?	How certain is it that the residual impacts will occur?	How significant are the residual impacts onsite?	How significant are the residual impacts of developing the Agua Caliente SEZ in the region (Sonoran Desert)?	Role in the ecosystem?	Other Considerations	Are potential residual impacts likely to warrant regional mitigation?
Ecology: Terrestrial Wildlife & Aquatic Biota	Yes.	Certain.	Very. Expect the loss of habitat for most general wildlife species over the entire developable area.	Decreasing trend in ecoregion.	Basic Component		Yes.
Ecology: Migratory Birds	Yes.	Certain. Linked to Vegetation & Riparian Areas.	Moderately High.	Decreasing trend in ecoregion.	Basic Component (along with other wildlife).	Need to consider prohibitions on take in the Migratory Bird Treaty Act MBTA, also direction in Executive Order 13186.	Yes.
Ecology: Plant Special Status Species (SSS)	No. No SSS plant species currently known on the SEZ.	Uncertain. If plant SSS are present, loss of habitat is certain, loss of individual plants is likely.	Negligible. If plant SSS are discovered during pre-disturbance surveys, expect the total loss of habitat in the developable area and loss of individual plants.		Basic Component (along with other vegetation).	Mitigation of SSS is required by BLM policy. Avoidance and minimization of impacts will be implemented.	No. Unless special status plant species are discovered during pre-disturbance surveys.
Ecology: Animal Special Status Species (SSS)	Yes.	Certain. Loss of habitat (for Conte's Thrasher, western burrowing owl, California leaf-nosed bat, and Pale Townsend's big-eared bat) is certain. Loss of animals is likely.	Very. Expect the total loss of habitat for animal SSS over the entire developable area.	Decreasing trend in ecoregion.	Basic Component (along with other wildlife).	Mitigation of SSS is required by BLM policy.	Yes.

Regional Mitigation Strategy for the Arizona SEZs

Table C-1. (Cont.)

Agua Caliente Solar Energy Zone Resource/ Issue	Residual or Unavoidable Impact?	How certain is it that the residual impacts will occur?	How significant are the residual impacts onsite?	How significant are the residual impacts of developing the Agua Caliente SEZ in the region (Sonoran Desert)?	Role in the ecosystem?	Other Considerations	Are potential residual impacts likely to warrant regional mitigation?
Environmental Justice	Maybe.		Depends on mitigation measures implemented on the basis of project-level NEPA.		Human Element		Maybe.
Hydrology: Surface Water	Yes.	Certain.	Moderate.	If storm water runoff is engineered appropriately to minimize modification of downstream resources, regional impacts can be avoided.	Basic Component		Maybe.
Hydrology: Water Quality & Groundwater	Maybe.	Uncertain (technology specific).	Uncertain (technology specific).	Uncertain (technology specific).	Basic Component		Maybe.
Military & Civilian Aviation	Maybe (with respect to military training routes).			Somewhat. Coordination with the military and possible height restrictions will address most impacts.	Human element		No. Through a combination of avoidance and design features, the potential impacts can be adequately mitigated onsite.

Table C-1. (Cont.)

Agua Caliente Solar Energy Zone Resource/ Issue	Residual or Unavoidable Impact?	How certain is it that the residual impacts will occur?	How significant are the residual impacts onsite?	How significant are the residual impacts of developing the Agua Caliente SEZ in the region (Sonoran Desert)?	Role in the ecosystem?	Other Considerations	Are potential residual impacts likely to warrant regional mitigation?
Native American Concerns	Yes.	Likely. Traditionally important plants will be destroyed and habitat for traditionally important animals will be lost. Important cultural resources will also be impacted.	Very—see Cultural, Wildlife, and SSS entries in this table.	Decreasing trend in ecoregion for wildlife and special status species also affects some Native American concerns	Human element	Consultation on project applications will determine whether regional mitigation may be warranted.	Unknown at this time.
Public Access & Recreation	Yes. Development may preclude current recreational activities that occur within the SEZ boundary.	Possible. Depends on mitigation measures implemented on the basis of project-level NEPA.	Moderate.	Low. Only small percentage of lands available in region is impacted.	Human element		Maybe.
Soils/Erosion	Yes.	Certain.	Very—expect disturbance to over the entire developable area.		Basic component		Yes.

Table C-1. (Cont.)

Agua Caliente Solar Energy Zone Resource/ Issue	Residual or Unavoidable Impact?	How certain is it that the residual impacts will occur?	How significant are the residual impacts onsite?	How significant are the residual impacts of developing the Agua Caliente SEZ in the region (Sonoran Desert)?	Role in the ecosystem?	Other Considerations	Are potential residual impacts likely to warrant regional mitigation?
Specially Designated Areas &Lands with Wilderness Characteristics	Maybe.	Possible. Some impacts to off-site user experience is expected, particularly for the Yuma East Undeveloped Special Recreation Management Area (SRMA), within the Gila River Valley Undeveloped SRMA, along the Juan Batista de Anza National Historic Trail, in the Sears Point Core portion of the Sears Point Area of Critical Environmental Concern (ACEC), and in the Gila River Terraces and Lower Gila Historic Trails ACEC.	Depends on whether locations of development within the SEZ impact key observation points; to be evaluated in project-level NEPA.	Depends on whether locations of development within the SEZ impact key observation points; to be evaluated in project-level NEPA.	Human element	Possible to minimize adverse visual impacts through on-site mitigation that reduces the degree of visual contrasts from new development.	Maybe.

Regional Mitigation Strategy for the Arizona SEZs

Table C-1. (Cont.)

Agua Caliente Solar Energy Zone Resource/ Issue	Residual or Unavoidable Impact?	How certain is it that the residual impacts will occur?	How significant are the residual impacts onsite?	How significant are the residual impacts of developing the Agua Caliente SEZ in the region (Sonoran Desert)?	Role in the ecosystem?	Other Considerations	Are potential residual impacts likely to warrant regional mitigation?
Visual	Yes.	Certain. Depending on the technology used, development in the SEZ may be visible in the community of Dateland, the Yuma East Undeveloped Special Recreation Management Area (SRMA), the Gila River Valley Undeveloped SRMA, along the Juan Batista de Anza National Historic Trail, in the Sears Point Core portion of the Sears Point Area of Critical Environmental Concern (ACEC), in the Gila River Terraces and Lower Gila Historic Trails ACEC, and along portions of Palomas Road and Interstate 8.	Depends on whether locations of development within the SEZ impact key observation points; to be evaluated in project-level NEPA. On-site mitigation and design would be implemented.	Depends on whether locations of development within the SEZ impact key observation points; to be evaluated in project-level NEPA.	Human element	Other resource mitigation that involves restoring the physical and biological integrity to the landscape may also mitigate visual resources as long as the visual design elements of form, line, color, and texture are factored into the restoration planning and design.	Maybe. Restoration or protection of intact ecosystems can also restore or protect visual resources.

Resources/Issues with no residual impacts: Climate Change, Invasive/Noxious Weeds, Riparian Areas, Lands & Realty, Livestock Grazing, Minerals, Paleontological, Socioeconomics, Transportation, Wild Horses & Burros.

Table C-2. Brenda Solar Energy Zone – Summary Table: Impacts That May Warrant Regional Compensatory Mitigation

Brenda Solar Energy Zone Resource/ Issue	Residual or Unavoidable Impact?	How certain is it that the residual impacts will occur?	How significant are the residual impacts onsite?	How significant are the residual impacts of developing the Brenda SEZ in the region (Sonoran Desert)?	Role in the ecosystem?	Other Considerations	Are potential residual impacts likely to warrant regional mitigation?
Acoustics	Maybe.	Possible. Depends on technology used and engineering controls.	Moderate. Construction-phase noise limited in duration; operation phase noise levels will be a permanent (30+ years) change.		Human Element		No. Generally impacts from solar development are expected to be temporary, localized, and readily mitigated.
Air Quality	Maybe (if site is graded).	Possible. Depends on whether the entire developable area (i.e., ~2,700 acres, 80% of the SEZ area) is graded			Human Element	Ambient PM Levels.	No. Impacts are not expected to result in noncompliance with National Air Quality Standards.
Cultural	Yes.	Certain.	Moderate. Results of archaeological survey of the entire SEZ identified significant (eligible) sites, but most will be avoided by creation of non-development area within the SEZ.	Moderate. Results of archaeological survey of the entire SEZ identified significant (eligible) sites, but most will be avoided by creation of non-development area within the SEZ.	Human Element	Avoidance preferred for significant resources. Adequate mitigation would be dependent on consultation and the resources and their relative significance in the region.	Maybe. Impacts warranting mitigation to be evaluated in consultation with the AZ State Historic Preservation Office (SHPO) and tribes.

Brenda Solar Energy Zone Resource/ Issue	Residual or Unavoidable Impact?	How certain is it that the residual impacts will occur?	How significant are the residual impacts onsite?	How significant are the residual impacts of developing the Brenda SEZ in the region (Sonoran Desert)?	Role in the ecosystem?	Other Considerations	Are potential residual impacts likely to warrant regional mitigation?
Ecology: Vegetation	Yes.	Certain.	Very. Expect the loss of all vegetation over the developable area of the SEZ, though mitigation may result in some remaining or replanted vegetation.	Decreasing trend in ecoregion.	Basic Component	Natural regeneration of native vegetation is slow in the Sonoran Desert.	Yes. Critical component of a functioning ecosystem.
Ecology: Riparian Areas	Maybe.	Possible. Depends on the degree of avoidance first then on engineering controls to address hydrologic impacts.		Decreasing trend in region, disproportionate historic impact to riparian areas.	Basic Component	Hydrologic impacts will affect riparian vegetation thus affect migratory birds, habitat for other wildlife—e.g., thermal cover for deer.	Yes.
Ecology: Invasive & Noxious Weeds	Maybe.	Possible. Depends on degree of vegetation disturbance and adequacy of Design Features.	Low.			Impacts will be minimized through development of a Weed Management Plan and use of weed-free seed to support re-vegetation efforts, control invasive species, and prevent increase in fires.	No.

Regional Mitigation Strategy for the Arizona SEZs

Brenda Solar Energy Zone Resource/Issue	Residual or Unavoidable Impact?	How certain is it that the residual impacts will occur?	How significant are the residual impacts onsite?	How significant are the residual impacts of developing the Brenda SEZ in the region (Sonoran Desert)?	Role in the ecosystem?	Other Considerations	Are potential residual impacts likely to warrant regional mitigation?
Ecology: Terrestrial Wildlife & Aquatic Biota	Yes.	Certain.	Very. Expect the loss of habitat for most general wildlife species over the entire developable area.	Decreasing trend in ecoregion.	Basic Component		Yes.
Ecology: Migratory Birds	Yes.	Probable. Linked to Vegetation & Riparian Areas.	Moderately High. Significance level will be re-evaluated when more monitoring data is available.	Decreasing trend in ecoregion.	Basic Component (along with other wildlife).	Need to consider prohibitions on take in the Migratory Bird Treaty Act MBTA, also direction in Executive Order 13186.	Yes.
Ecology: Plant Special Status Species (SSS)	No. No SSS plant species currently known on the SEZ.	Uncertain. If plant SSS are present, loss of habitat is certain, loss of individual plants is likely.	Negligible. If plant SSS are discovered during pre-disturbance surveys, expect the total loss of habitat in the developable area and loss of individual plants.		Basic Component (along with other vegetation).	Mitigation of SSS is required by BLM policy. If identified avoidance and minimization of impacts would be implemented.	No. Unless special status plant species are discovered during pre-disturbance surveys.
Ecology: Animal Special Status Species (SSS)	Yes.	Loss of habitat (for Sonoran desert tortoise, Western burrowing owl, California leaf-nosed bat, and Pale Townsend's big-eared bat)) is certain. Loss of animals is likely.	Very. Expect the total loss of habitat for animal SSS over the entire developable area.	Decreasing trend in ecoregion.	Basic Component (along with other wildlife).	Mitigation of SSS is required by BLM policy.	Yes.

Regional Mitigation Strategy for the Arizona SEZs

Brenda Solar Energy Zone/ Issue	Residual or Unavoidable Impact?	How certain is it that the residual impacts will occur?	How significant are the residual impacts onsite?	How significant are the residual impacts of developing the Brenda SEZ in the region (Sonoran Desert)?	Role in the ecosystem?	Other Considerations	Are potential residual impacts likely to warrant regional mitigation?
Environmental Justice	Maybe.		Depends on mitigation measures implemented on the basis of project-level NEPA.		Human Element		Maybe.
Hydrology: Surface Water	Yes.	Certain.	Moderate.	If storm water runoff is engineered appropriately to minimize modification of downstream resources, regional impacts can be avoided.	Basic Component		Maybe.
Military & Civilian Aviation	Maybe (with respect to MTRs).			Somewhat. Coordination with the military and possible height restrictions will address most impacts.	Human element		No. Through a combination of avoidance and design features, the potential impacts can be adequately mitigated onsite.
Native American Concerns	Yes.	Likely. Traditionally important plants will be destroyed and habitat for traditionally important animals will be lost. Important cultural resources will also be impacted.	Very—see Cultural, Wildlife, and SSS entries in this table.	Decreasing trend in ecoregion for wildlife and special status species also affects some Native American concerns.	Human element	Consultation on project applications will determine whether regional mitigation for may be warranted.	Unknown at this time.

Regional Mitigation Strategy for the Arizona SEZs

Brenda Solar Energy Zone Resource/ Issue	Residual or Unavoidable Impact?	How certain is it that the residual impacts will occur?	How significant are the residual impacts onsite?	How significant are the residual impacts of developing the Brenda SEZ in the region (Sonoran Desert)?	Role in the ecosystem?	Other Considerations	Are potential residual impacts likely to warrant regional mitigation?
Public Access & Recreation	Yes. Development will preclude current recreational activities that occur within the SEZ boundary.	Possible. Depends on mitigation measures implemented on the basis of project-level NEPA.	Moderate.	Low. Only small percentage of lands available in region is impacted.	Human element		Maybe.
Soils/Erosion	Yes.	Certain.	Very— expect disturbance to cover the entire developable area.		Basic component		Yes.
Specially Designated Areas &Lands with Wilderness Characteristics	Yes.	Possible. Some impacts to off-site user experience are expected. Minimal visual impacts are expected.	Depends on whether locations of development within the SEZ impact key observation points; to be evaluated in project-level NEPA.	Depends on whether locations of development within the SEZ impact key observation points; to be evaluated in project-level NEPA.	Human element	Possible to minimize adverse visual impacts through on-site mitigation that reduces the degree of visual contrasts from new development.	No.

Regional Mitigation Strategy for the Arizona SEZs

Brenda Solar Energy Zone Resource/ Issue	Residual or Unavoidable Impact?	How certain is it that the residual impacts will occur?	How significant are the residual impacts onsite?	How significant are the residual impacts of developing the Brenda SEZ in the region (Sonoran Desert)?	Role in the ecosystem?	Other Considerations	Are potential residual impacts likely to warrant regional mitigation?
Visual	Yes.	Certain. Depending on the technology used, development in the SEZ will be readily visible in the communities of Vicksburg and Brenda, in the Plomosa SRMA, La Posa Destination SRMA, and along U.S. Highway 60 and Interstate 10.	Depends on whether locations of development within the SEZ impact key observation points; to be evaluated in project-level NEPA. On-site mitigation and design would be implemented	Depends on whether locations of development within the SEZ impact key observation points; to be evaluated in project-level NEPA.	Human element	Other resource mitigation that involves restoring the physical and biological integrity to the landscape may also mitigate visual resources as long as the visual design elements of form, line, color, and texture are factored into the restoration planning and design.	Maybe. Restoration or protection of intact ecosystems can also restore or protect visual resources.

Resources/Issues with no residual impacts: Climate Change, Lands & Realty, Livestock Grazing, Minerals, Paleontological, Socioeconomics, Transportation, Wild Horses & Burros.

Table C-3. Gillespie Solar Energy Zone – Summary Table: Impacts That May Warrant Regional Compensatory Mitigation

Gillespie Solar Energy Zone Resource/ Issue	Residual or Unavoidable Impact?	How certain is it that the residual impacts will occur?	How significant are the residual impacts onsite?	How significant are the residual impacts of developing the Gillespie SEZ in the region (Sonoran Desert)?	Role in the ecosystem?	Other Considerations	Are potential residual impacts likely to warrant regional mitigation?
Acoustics	Maybe.	Possible. Depends on technology used and engineering controls.	Moderate. Construction-phase noise limited in duration; operation phase noise levels will be a permanent (30+ years) change.		Human Element		No. Generally impacts from solar development are expected to be temporary, localized, and readily mitigated.
Air Quality	Maybe (if site is graded).	Possible. Depends on whether the entire developable area (i.e., 1,785 acres, 80% of the developable SEZ area) is graded.			Human Element	Ambient PM Levels.	No.
Cultural	Yes.	Possible.	Low.	Low.	Human Element	Avoidance preferred for significant resources. Adequate mitigation would be dependent on consultation and the resources and their relative significance in the region.	Maybe. Impacts warranting mitigation to be evaluated in consultation with the AZ State Historic Preservation Office (SHPO) and tribes.

Regional Mitigation Strategy for the Arizona SEZs

Gillespie Solar Energy Zone Resource/ Issue	Residual or Unavoidable Impact?	How certain is it that the residual impacts will occur?	How significant are the residual impacts onsite?	How significant are the residual impacts of developing the Gillespie SEZ in the region (Sonoran Desert)?	Role in the ecosystem?	Other Considerations	Are potential residual impacts likely to warrant regional mitigation?
Ecology: Vegetation	Yes.	Certain.	Very. Expect the loss of all vegetation over the developable area of the SEZ, though mitigation may result in some remaining or replanted vegetation.	Decreasing trend in ecoregion.	Basic Component	Natural regeneration of native vegetation is slow in the Sonoran Desert.	Yes. Critical component of a functioning ecosystem.
Ecology: Riparian Areas	Maybe.	Possible. Depends on the degree of avoidance first, then on engineering controls to address hydrologic impacts.		Decreasing trend in region, disproportionate historic impact to riparian areas.	Basic Component	Hydrologic impacts will affect riparian vegetation, thus affect migratory birds, habitat for other wildlife—e.g., thermal cover for deer.	Yes.
Ecology: Invasive & Noxious Weeds	Maybe.	Possible. Depends on degree of vegetation disturbance and adequacy of Design Features.	Low.			Impacts will be minimized through development of a Weed Management Plan and use of weed-free seed to support re-vegetation efforts, control invasive species, and prevent increase in fires.	No.

Regional Mitigation Strategy for the Arizona SEZs

Gillespie Solar Energy Zone Resource/ Issue	Residual or Unavoidable Impact?	How certain is it that the residual impacts will occur?	How significant are the residual impacts onsite?	How significant are the residual impacts of developing the Gillespie SEZ in the region (Sonoran Desert)?	Role in the ecosystem?	Other Considerations	Are potential residual impacts likely to warrant regional mitigation?
Ecology: Terrestrial Wildlife & Aquatic Biota	Yes.	Certain.	Very. Expect the loss of habitat for most general wildlife species over the entire developable area.	Decreasing trend in ecoregion.	Basic Component		Yes.
Ecology: Migratory Birds	Yes.	Probable. Linked to Vegetation & Riparian Areas.	Moderate to High.	Decreasing trend in ecoregion.	Basic Component (along with other wildlife).	Need to consider prohibitions on take in the Migratory Bird Treaty Act MBTA, also direction in Executive Order 13186.	Yes.
Ecology: Plant Special Status Species (SSS)	No. No SSS plant species currently known on the SEZ.	Uncertain. If plant SSS are present, loss of habitat is certain, loss of individual plants is likely.	Negligible. If plant SSS are discovered during pre-disturbance surveys, expect the total loss of habitat in the developable area and loss of individual plants.		Basic Component (along with other vegetation).	Mitigation of SSS is required by BLM policy. If identified avoidance and minimization of impacts would be implemented.	No. Unless special status plant species are discovered during pre-disturbance surveys.
Ecology: Animal Special Status Species (SSS)	Yes.	Certain. Loss of habitat (for Sonoran Desert tortoise, Western burrowing owl, California leaf-nosed bat, and possibly Mexican rosy boa) is certain. Loss of animals is likely.	Very. Expect the total loss of habitat for animal SSS over the entire developable area.	Decreasing trend in ecoregion.	Basic Component (along with other wildlife).	Mitigation of SSS is required by BLM policy.	Yes.

Regional Mitigation Strategy for the Arizona SEZs

Gillespie Solar Energy Zone Resource/ Issue	Residual or Unavoidable Impact?	How certain is it that the residual impacts will occur?	How significant are the residual impacts onsite?	How significant are the residual impacts of developing the Gillespie SEZ in the region (Sonoran Desert)?	Role in the ecosystem?	Other Considerations	Are potential residual impacts likely to warrant regional mitigation?
Hydrology: Surface Water	Yes.	Certain.	Moderate.	If storm water runoff is engineered appropriately to minimize modification of downstream resources, regional impacts can be avoided.	Basic Component		Maybe.
Livestock Grazing	Maybe.	Low.	Depends on mitigation measures implemented on the basis of project-level NEPA.		Land Use	Permittees would be compensated for lost or otherwise impacted range improvements.	Maybe. Through a combination of avoidance and design features, the potential impacts can likely be adequately mitigated onsite.
Military & Civilian Aviation	Maybe (with respect to military training routes).			Somewhat. Coordination with the military and possible height restrictions will address most impacts.	Human element		No. Through a combination of avoidance and design features, the potential impacts can be adequately mitigated onsite.

Regional Mitigation Strategy for the Arizona SEZs

Gillespie Solar Energy Zone Resource/ Issue	Residual or Unavoidable Impact?	How certain is it that the residual impacts will occur?	How significant are the residual impacts onsite?	How significant are the residual impacts of developing the Gillespie SEZ in the region (Sonoran Desert)?	Role in the ecosystem?	Other Considerations	Are potential residual impacts likely to warrant regional mitigation?
Native American Concerns	Yes.	Likely that traditionally-important plants will be destroyed and that habitat for traditionally-important animals will be lost. Unknown for cultural resources until Class III cultural inventories are completed.	Very—see Wildlife and SSS entries in this table.	Decreasing trend in ecoregion for wildlife and special status species also affects some Native American concerns.	Human element	Consultation on project applications will determine whether regional mitigation for may be warranted.	Unknown at this time.
Public Access & Recreation	Maybe. Development may preclude current recreational activities that occur within the SEZ boundary.	Possible. Depends on mitigation measures implemented on the basis of project-level NEPA.	Low. Relatively little recreation currently occurs in the SEZ.	Low. Only small percentage of lands available in region is impacted.	Human element		Maybe.
Soils/Erosion	Yes.	Certain.	Very—expect disturbance to over the entire developable area.		Basic component		Yes.

C-20

Regional Mitigation Strategy for the Arizona SEZs

Gillespie Solar Energy Zone Resource/ Issue	Residual or Unavoidable Impact?	How certain is it that the residual impacts will occur?	How significant are the residual impacts onsite?	How significant are the residual impacts of developing the Gillespie SEZ in the region (Sonoran Desert)?	Role in the ecosystem?	Other Considerations	Are potential residual impacts likely to warrant regional mitigation?
Specially Designated Areas & Lands with Wilderness Characteristics	Yes.	Possible. Some impacts to off-site user experience are expected, particularly for Signal Mountain WA, Woolsey Peak WA, Saddle Mountain SRMA, and the proposed backcountry byway.	Depends on whether locations of development within the SEZ impact key observation points; to be evaluated in project-level NEPA.	Depends on whether locations of development within the SEZ impact key observation points; to be evaluated in project-level NEPA.	Human element	Possible to minimize adverse visual impacts through on-site mitigation that reduces the degree of visual contrasts from new development.	Maybe.

Regional Mitigation Strategy for the Arizona SEZs

Gillespie Solar Energy Zone Resource/ Issue	Residual or Unavoidable Impact?	How certain is it that the residual impacts will occur?	How significant are the residual impacts onsite?	How significant are the residual impacts of developing the Gillespie SEZ in the region (Sonoran Desert)?	Role in the ecosystem?	Other Considerations	Are potential residual impacts likely to warrant regional mitigation?
Visual	Yes.	Certain. Development in the SEZ will be readily visible in the community of Arlington, on Agua Caliente Road (proposed backcountry byway), and from several specially designated areas.	Depends on whether locations of development within the SEZ impact key observation points; to be evaluated in project-level NEPA. On-site mitigation and design would be implemented.	Depends on whether locations of development within the SEZ impact key observation points; to be evaluated in project-level NEPA.	Human element	The SEZ Visual Resource Management (VRM) class is IV and allows for development. Other resource mitigation that involves restoring the physical and biological integrity to the landscape may also mitigate visual resources as long as the visual design elements of form, line, color, and texture are factored into the restoration planning and design.	Maybe. Restoration or protection of intact ecosystems can also improve scenic quality.

Resources/Issues with no residual impacts: Climate Change, Environmental Justice, Lands & Realty, Minerals, Paleontological, Socioeconomics, Transportation, Wild Horses & Burros.

APPENDIX D

**BLM Screening of Candidate Regional Mitigation Sites
for the Arizona Solar Energy Zones**

This page intentionally left blank

APPENDIX D

**BLM Screening of Candidate Regional Mitigation Sites
for the Arizona Solar Energy Zones**

The BLM is currently considering many potential mitigation sites and actions as listed in Section 2.8 of the SRMS. These sites were nominated by stakeholders and the BLM. This appendix includes two maps presenting all of the candidate site locations (Figures D-1 and D-2). All of the sites listed on these maps were evaluated by the BLM in the matrix that follows.

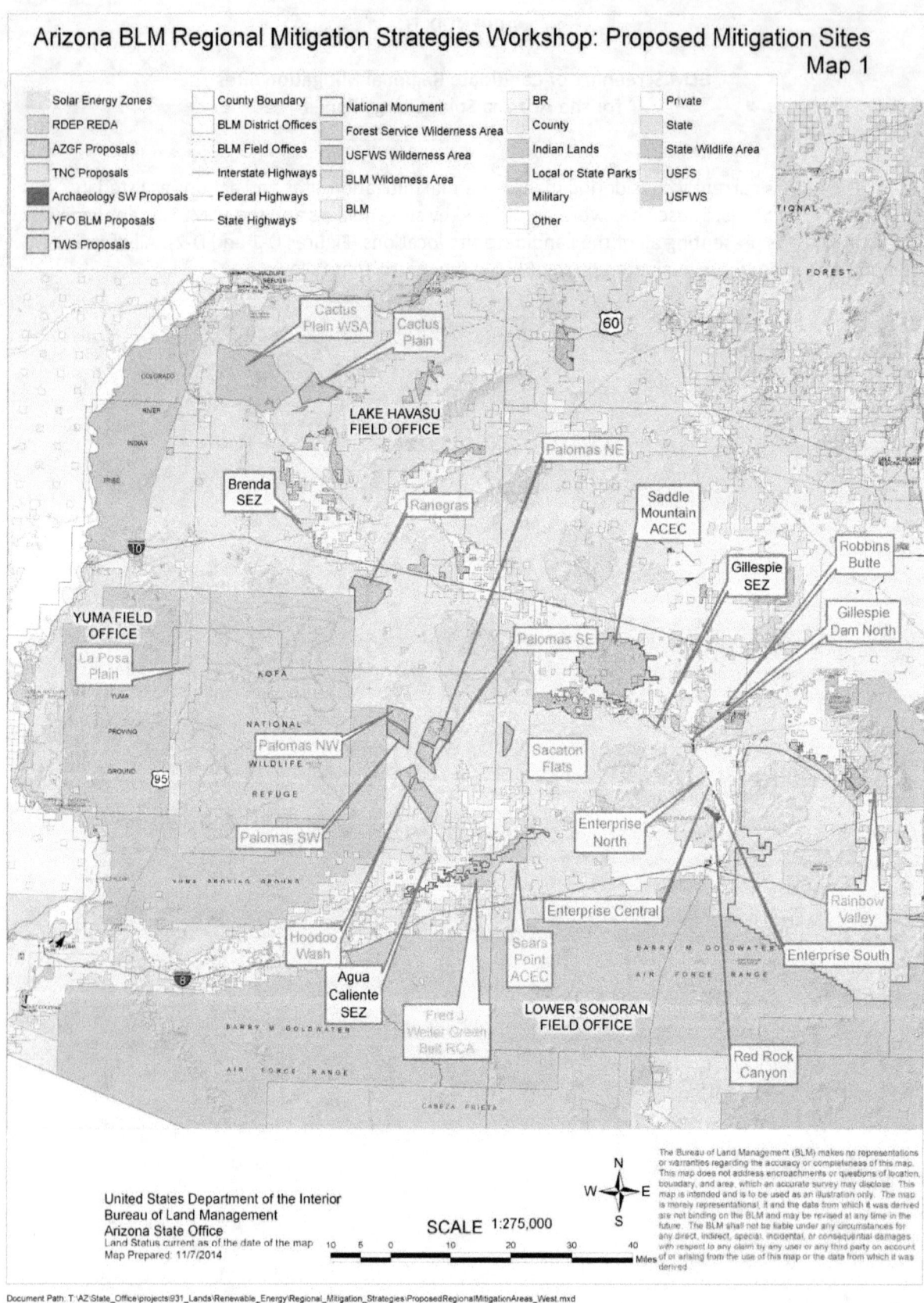

Figure D-1. Recommended Candidate Regional Mitigation Sites for Arizona SEZs, western sites

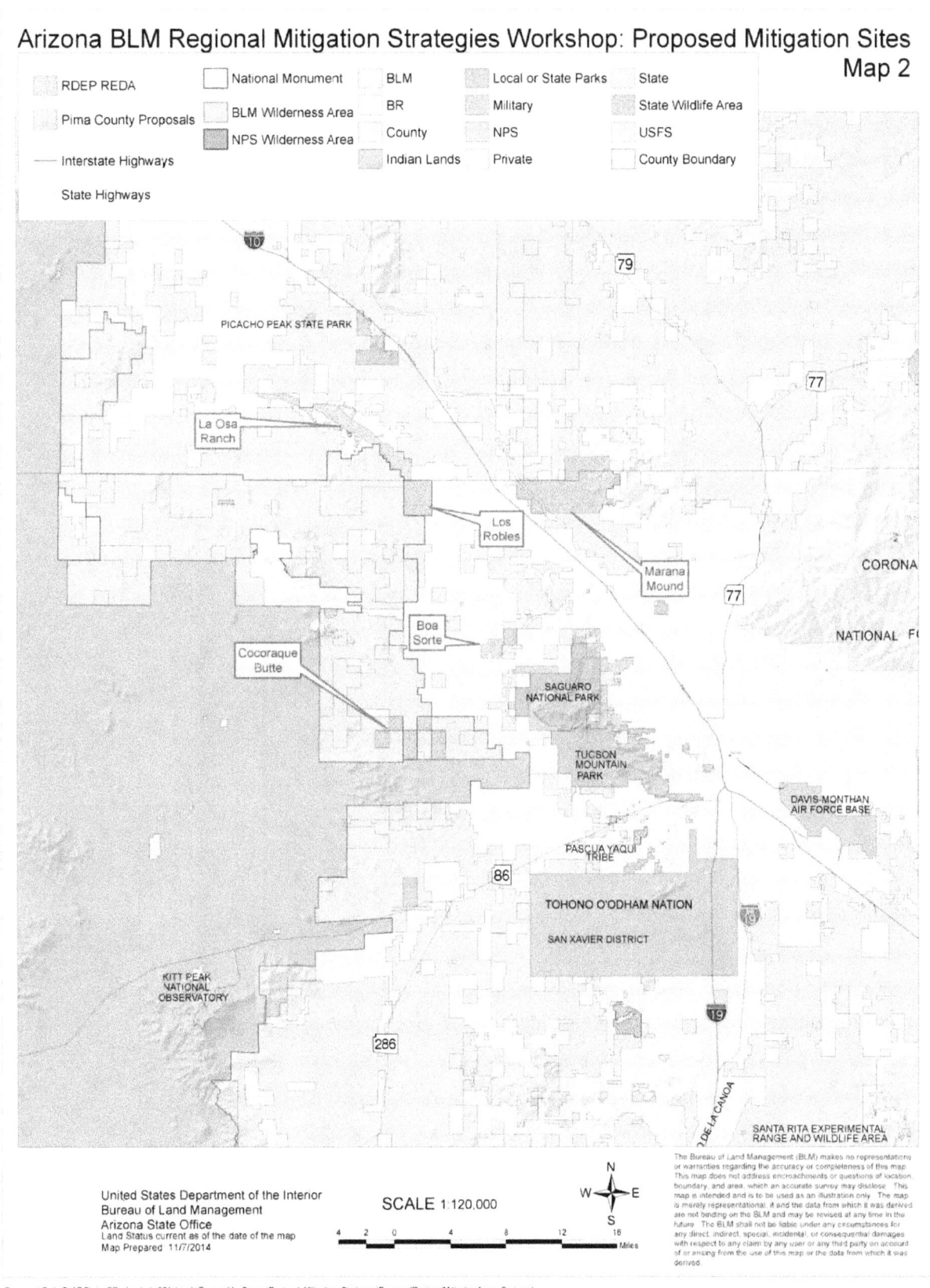

Figure D-2. Recommended Candidate Regional Mitigation Sites for Arizona SEZs, eastern sites

This page intentionally left blank

Table D-1. BLM Matrix for Candidate Regional Mitigation Sites for Arizona SEZs

Criteria	SEZs Being Evaluated			Candidate Sites			
	Brenda	Gillespie	Agua Caliente	Cactus Plain WSA (TNC)	La Posa Plain (TNC)	Rainbow Valley (TNC)	
SITE CHARACTERISTICS							
1. Total area of site (acres)	3,348	2,618	2,550	58,928	64,880	24,439	
BLM acres	3,348	2,618	2,550	58,928	64,073	21,644	
Private acres					807	2,795	
State Trust acres							
2. Sources of data for the site.	Solar PEIS	Solar PEIS	Arizona RDEP	USGS Protected Areas Database of the United States, TNC Ecoregional rollup, BLM REA, Lake Havasu RMP, STATSGO, SURGO for soil stability; wildlife linkages	USGS Protected Areas Database of the United States, TNC Ecoregional rollup, BLM REA, Yuma RMP, STATSGO, SURGO for soil stability; wildlife linkages	USGS Protected Areas Database of the United States, TNC Ecoregional rollup, BLM REA, Lower Sonoran RMP, STATSGO, SURGO for soil stability; wildlife linkages	
3. Mitigates for all or most identified residual impacts that may warrant compensatory mitigation. √ for yes (1 point) or X for no (-2 points); Include justification.	Vegetation, wildlife, migratory birds, SSS Animals, hydrology, soils Cultural, visual resources	Vegetation, wildlife, migratory birds, SSS Animals, hydrology, soils Cultural, specially designated areas, visual resources	Vegetation, wildlife, migratory birds, SSS Plants, SSS Animals, hydrology, soils Cultural, specially designated areas, visual resources, recreation	√ Ecological Resources: SSS Animals	√ Ecological Resources: SSS Animals	√ Ecological Resources: SSS Animals	
4. Mitigation action (restoration/enhancement, acquisition, withdrawal, special designation, etc.).				Restoration; closure and revegetation of unauthorized roads.	Acquisition; restoration removal of barriers; tortoise fencing; wildlife crossing structures.	Acquisition; removal of barriers; restoration; wildlife crossing structures, wildlife fencing and tortoise fencing.	
5. Site and its proposed actions meet regional conservation/mitigation goals and objectives. √ for yes (1 point) or X for no (-2 points).				√	√	√	
Justification.				Preserves and restores creosote/bursage habitat; enhances visual quality; protects BLM sensitive species.	Preserves and restores creosote/bursage habitat; protects BLM sensitive species.	Protects tortoise habitat; preserves and restores creosote/bursage habitat; protects BLM sensitive species; protects cultural resources.	

Regional Mitigation Strategy for the Arizona SEZs

Table D.1. (Cont.)

Criteria	SEZs Being Evaluated			Candidate Sites		
	Brenda	Gillespie	Agua Caliente	Cactus Plain WSA (TNC)	La Posa Plain (TNC)	Rainbow Valley (TNC)
6. Proposed Mitigation Action and location Consistent with the Resource Management Plan. √ for yes (1 point) or X for no (-2 points); Include justification.				√	√	√
7. Same HUC 4 watershed. Specify watershed.	1503-Lower Colorado River below Lake Mead	1507-Lower Gila River	1507-Lower Gila River	√ Lower Colorado	√ Lower Colorado	√ Lower Gila
8. VRI Class and acres associated with each class.	Class IV 3,345 acres	Class III 2,618 acres	Class III 2,543 acres	Class IV 11,643 acres	Class III 63,983 acres	Class I: 3 acres; Class II: 1,046; Class III: 10,131; Class IV: 10,474 acres
9. Similar landscape value, ecological functionality, biological value, species, habitat types, and/or natural features. Score based on responses to criteria 9a and 9b. √ for yes (1 point) or X for no (-2 points).				√	√	√
9a. Current terrestrial landscape intactness score (use Sonoran Desert Rapid Ecological Assessment Data) and acres associated with each intactness category[27] .	Very Low – 4 acres; Low 57 acres; Mod Low 363 acres; Mod High 1,895 acres; High 1,027 acres	Very Low – 17 acres; Low 174 acres; Mod Low 461 acres; Mod High 1,419 acres; Very High 258 acres	Very Low – 2,214 acres; Low 44 acres; Mod Low 189 acres; Mod High 96 acres;	Very Low: 436 acres; Low: 17,417 acres; Mod Low: 21,465 acres; Mod High: 14,980 acres; High: 4,463 acres Very High: 99 acres	Very Low: 1,236 acres; Low: 9,791 acres; Mod Low: 21,465 acres; Mod High: 14,980 acres; High: 4,463 acres Very High: 99 acres	Very Low: 1,284 acres; Low: 1,315 acres; Mod Low: 4,133 acres; Mod High: 7,491 acres; High: 3,112 acres Very High: 7,118 acres
9b. Dominant vegetation communities.	Sonoran-Mojave Creosotebush-White Bursage Desert Scrub (91%) Sonoran Paloverde-Mixed Cacti Desert Scrub (9%)	Sonoran-Mojave Creosotebush-White Bursage Desert Scrub (95%) Sonoran Paloverde-Mixed Cacti Desert Scrub (5.4-3%)	Sonoran-Mojave Creosotebush-White Bursage Desert Scrub (96%) Introduced Vegetation (2.8%)	Sonoran-Mojave Creosotebush-White Bursage Desert Scrub (85%) Sonoran Paloverde-Mixed Cacti Desert Scrub (14%)	Sonoran-Mojave Creosotebush-White Bursage Desert Scrub (80%) Sonoran Paloverde-Mixed Cacti Desert Scrub (20%)	Sonoran-Mojave Creosotebush-White Bursage Desert Scrub (81%) Sonoran Paloverde-Mixed Cacti Desert Scrub (16%)
10. In SEZ Ecoregion. Specify ecoregion. √ for yes (1 point) or X for no (-2 points).	Sonoran Desert	Sonoran Desert	Sonoran Desert	√	√	√

[27] Intactness Categories: Very Low (-1.0 – -0.75); Low (-0.75 – -0.5); Mod Low (-0.5 – 0.0); Mod High (0.0 – 0.5); High (0.5 – 0.75) Very High (0.75 – 1.0).

Regional Mitigation Strategy for the Arizona SEZs

Table D.1. (Cont.)

Criteria	SEZs Being Evaluated			Candidate Sites		
	Brenda	Gillespie	Agua Caliente	Cactus Plain WSA (TNC)	La Posa Plain (TNC)	Rainbow Valley (TNC)
11. In SEZ ecological subregion. √ for yes (1 point) or X for no (-2 points).	Colorado Desert-EPA: 81j	Colorado Desert-EPA: 81j	Colorado Desert-EPA: 81j	√ (Colorado Desert - EPA: 81j, 81d)	√ (Colorado Desert - EPA: 81j)	√ (Colorado Desert - EPA: 81j; Arizona upland- EPA: 81k)
12. Provides adequate geographic extent. Depending on whether site provides area for mitigation at least as large as the entire developable area of the SEZ. √ for yes (1 point) or X for no (-2 points).				√	√	√
FEASIBILITY						
13. Feasibility of action Justification of feasibility score: Scores for 13a through 13 e were provided by stakeholders. BLM used these scores as well as their knowledge of the sites and actions.				5 Closure and revegetation of roads is straightforward and low cost.	4 Removal of barriers is straightforward and low cost. Wildlife crossing structures and tortoise fencing are established tools with many examples. Land acquisition can be complex, but is not critical to success at this site.	4 Removal of barriers is straightforward and low cost. Wildlife crossing structures are an established tool with many examples. Land acquisition can be complex, but is not critical to success at this site. Restoration of ag fields can range from passive and cheap to very active and moderately expensive, depending on goals and resources available.

Regional Mitigation Strategy for the Arizona SEZs

Table D.1. (Cont.)

Criteria	SEZs Being Evaluated			Candidate Sites		
	Brenda	Gillespie	Agua Caliente	Cactus Plain WSA (TNC)	La Posa Plain (TNC)	Rainbow Valley (TNC)
13a. What level of documentation is available to demonstrate effectiveness of mitigation action? Use scale of 1 (little to no documentation) to 5 (well documented).				5	5	5
Justification.				Road closure and revegetation is a widespread practice with good results.	Bighorn need for cross-basin connectivity without barriers (Monson and Sumner 1980). Tortoise need for cross-basin connectivity without barriers (Edwards et al. 2004; Conservation Genetics 5.4: 485-499).	Bighorn need for cross-basin connectivity without barriers: Monson and Sumner 1980. Tortoise need for cross-basin connectivity without barriers: Edwards et al. 2004. Conservation Genetics 5.4: 485–499. AZ Wildlife Linkages Assessment (Nordhaugen et al. 2006); The Maricopa County Connectivity Assessment: Report on Stakeholder Input (January 2012); and Beier et al. (2008) Arizona Missing Linkages Gila Bend – Sierra Estrella Linkage Design.

Regional Mitigation Strategy for the Arizona SEZs

Table D.1. (Cont.)

Criteria	SEZs Being Evaluated			Candidate Sites		
	Brenda	Gillespie	Agua Caliente	Cactus Plain WSA (TNC)	La Posa Plain (TNC)	Rainbow Valley (TNC)
13b. [28] Based on action required (e.g., restoration, BLM land management action, land acquisition, Congressional action), how difficult will implementation be? Use scale of 1 (difficult) to 5 (relatively easy).				5	3	4
Justification.						
13c. Time frame needed to establish site as mitigation location (estimated years).				1 years	2 years	2 years
13d. Time frame for achieving mitigation goals and objectives from implementation (estimated years).				2 years	5 years	5 years
13e. Cost estimate.				$5,000	$2,700,000 to $4,900,000	$35,000 to $9,200,000
EFFECTIVENESS / ADDITIONALITY						
14. Effectiveness and Additionality				3	4	5
Justification of effectiveness and additionality score: Scores for 14a through 14c were provided by stakeholders. BLM used these scores as well as their knowledge of the sites and actions.				This site is a wilderness study area with high intactness. Work in the area will be an improvement, but incremental.		
14a. [29] To what extent can the full spectrum of mitigation goals/objectives be met simultaneously? Use scale of 0 (low) to 5 (high).				1	3	5
Justification.						

---

[28] Rate the mitigation action, based on the following scale: restoration/enhancement actions (score of 1–3); Congressional actions (score of 1). Ratings should be adjusted on the basis of factors such as cost of the action; time and effort requirements; public and/or BLM support for or opposition to action; and, for land acquisitions, willingness of seller. restoration/enhancement actions (score of 5); BLM planning decisions (score of 3–4); land acquisition actions (score

[29] Rate the extent to which the mitigation desired outcomes can be met simultaneously through mitigation actions at the site, based on the following scale: all (100%) of the goals and objectives can be met (score of 5); 75–99% can be met (score of 4); 50–74% (score of 3); 25–49% can be met (score of 2); less than 25% can be met (score of 1); none of the goals/objectives can be met (score of 0).

Regional Mitigation Strategy for the Arizona SEZs

Table D.1. (Cont.)

Criteria	SEZs Being Evaluated			Candidate Sites		
	Brenda	Gillespie	Agua Caliente	Cactus Plain WSA (TNC)	La Posa Plain (TNC)	Rainbow Valley (TNC)
14b. [30] How effective will the mitigation be in the context of achieving mitigation goals/objectives for conserving/restoring ecosystem intactness? Use scale of 1 (low) to 5 (high).				4	4	5
Justification.						
14c. Mitigation consists of actions that would not otherwise be undertaken by BLM.				Maybe	Yes	
RISK						
15. Risk of action(s)				3	4	4
Justification of risk score: Scores for 15a through 15b were provided by stakeholders. BLM used these scores as well as their knowledge of the sites and actions.				Risk of illegal use and reopening of closed roads.	Low risk of development but difficult to fund.	Some risk of development pressure (Sonoran Valley Parkway and Sonoran Solar nearby). However, located between wilderness and National Monument.
15a. What are the constraints or threats to success?				Vandalism and off-road drivers reopening closed roads.	Expense of wildlife crossing structures.	Cost of land acquisition. Conditions of new roadway construction. Expense of wildlife crossing structures.

[30] Rate the effectiveness of the mitigation actions at the site in terms of achieving mitigation goals/objectives, based on the following scale: highly effective (score of 5); moderately effective (scores of 2–4), and minimally effective (score of 1).

Regional Mitigation Strategy for the Arizona SEZs

Table D.1. (Cont.)

Criteria	SEZs Being Evaluated			Candidate Sites		
	Brenda	Gillespie	Agua Caliente	Cactus Plain WSA (TNC)	La Posa Plain (TNC)	Rainbow Valley (TNC)
15b. What are surrounding land uses that will impact mitigation success (e.g., proximity to expanding urban areas, pressures on region for recreational land use, excessive groundwater withdrawal and drawdown conditions that could affect resources on the mitigation site)?						The City of Goodyear proposes to develop on either side of the Rainbow Valley site so urbanization is a possibility. Also there is an alternative segment for the I-11 corridor which may cross the site in the future. Neither precludes BLM management of the site; however, they will both have indirect effects on the site that will require mitigation designs to avoid and/or minimize.
DURABILITY						
16. Durability of action(s)				4	3	3
Justification of durability score: Scores for 16a and 16b were provided by stakeholders. BLM used these scores as well as their knowledge of the sites and actions.				Protected, but risk from continued illegal use lowers managerial durability.	Has some designations (SRMA, WHA) in RMP and some restrictions on land use authorizations.	
16a. [31] How durable would the mitigation be from a time frame and management perspective? Use scale of 1 (low) to 5 (high).				5	4	4
Justification.						

---

[31] Rate the temporal and managerial durability of the mitigation action, based on the following scale: Congressionally protected lands would be very durable (score of 5); other federally administered lands specifically designated in land use plans or withdrawn by public land order would be moderately to very durable (score of 4–5); federally administered lands without any special designation but with enforcement oversight would have limited durability (score of 2); lands without special designation or enforcement oversight would not be very durable (score of 1).

Regional Mitigation Strategy for the Arizona SEZs

Table D.1. (Cont.)

Criteria	SEZs Being Evaluated			Candidate Sites		
	Brenda	Gillespie	Agua Caliente	Cactus Plain WSA (TNC)	La Posa Plain (TNC)	Rainbow Valley (TNC)
16b. Are there potential effects of future climate change[32]?				High	Moderate	High
PRELIMINARY SCORING Calculate score by summing the entries in blue-shaded cells. Scores are calculated based on entries in blue-shaded cells as follows: all scaled values (i.e., ratings from 1 to 5) are summed; 1 point is added for each V; 2 points are deleted for each X.				22	22	23
ADDITIONAL CONSIDERATIONS						
17. Presence of unique and/or valuable resources or features. (Up to 3 additional points for unique and/or valuable resources or features present at the candidate site, in 17a through 17h.)				2	1	3
17a. Perennial, protected sources of water.	NA	NA	NA	NA	NA	NA
17b. Unique species assemblages.				Sand dune community.		High diversity of SCGN bird and reptile species.
17c. AZGFD Species of Greatest Conservation Need (distribution models)/ Heritage Data Management System species (occurrence data).	24/0	26/0	31/0	27/7	26/3	52/1
17d. BLM categorized desert tortoise habitat.						Category 1 and 2 (on the edge).

---

32 Climate change categories are from the Sonoran Desert Rapid Ecoregional Assessment future climate change model (BLM 2011).

Regional Mitigation Strategy for the Arizona SEZs

Table D.1. (Cont.)

Criteria	SEZs Being Evaluated			Candidate Sites		
	Brenda	Gillespie	Agua Caliente	Cactus Plain WSA (TNC)	La Posa Plain (TNC)	Rainbow Valley (TNC)
17e. T&E species or critical habitat and/or BLM sensitive species	No T&E species or critical habitat. BLM sensitive species: Western burrowing owl, CA leaf-nosed bat, Sonoran desert tortoise, Pale Townsend's big-eared bat.	No T&E species or critical habitat. BLM sensitive species: Mexican rosy boa, Western burrowing owl, CA leaf-nosed bat, Sonoran desert tortoise.	No T&E species or critical habitat. BLM sensitive species: Western burrowing owl, CA leaf-nosed bat, Le Conte's thrasher, Pale Townsend's big-eared bat.	No T&E species or critical habitat. BLM sensitive species: Western burrowing owl, Gilded Flicker, Peregrine Falcon, Le Conte's Thrasher, Pale Townsend's big-eared bat, Spotted Bat, Greater Western Mastiff Bat, CA leaf-nosed bat, Cave Myotis, Mohave Fringe-toed Lizard.	No T&E species or critical habitat. BLM sensitive species: Golden Eagle, Gilded Flicker, Le Conte's Thrasher, Pale Townsend's big-eared bat, Spotted Bat, Greater Western Mastiff Bat, CA leaf-nosed bat, Arizona Myotis, Cave Myotis.	No T&E species or critical habitat. BLM sensitive species: Golden Eagle, Western burrowing owl, Gilded Flicker, Cactus Ferruginous Pygmy-Owl, Bald Eagle, Le Conte's Thrasher, Pale Townsend's big-eared bat, Spotted Bat, Greater Western Mastiff Bat, CA leaf-nosed bat, Cave Myotis.
17f. Desert washes (miles) or ephemeral playas (acres)				desert washes (1.9 miles)	desert washes (13.8 miles)	desert washes (4.7 miles)
17g. Known highly significant and unique cultural resources						National Historic Trail
17h. Other						Wildlife movement connectivity for bighorn, tortoise between SDNM and Sierra Estrella Wilderness.
18. Links two or more protected areas √ for yes (1 point) or 0 for no (no score adjustment); include justification.				0 Adjacent to East Cactus Plains and Gibraltar Mountain Wilderness.	√ Between Kofa NWR (protected) and Yuma Proving Ground (long-term withdrawal; existing Integrated National Resources Management Plan).	√ Between Sonoran Desert National Monument (protected) and Sierra Estrella Mountains (land use exclusion area).
COMBINED SCORE Add preliminary score to the additional consideration criteria in the blue-shaded cells. Scores are calculated based on entries in blue-shaded cells as follows: scaled values (i.e., ratings from 1 to 3) are summed; 1 point is added for each √.	NA	NA	NA	24	24	27

Regional Mitigation Strategy for the Arizona SEZs

Criteria	Candidate Sites				
	Ranegras Plain (AZGFD)	Sacaton Flats (AZGFD)	Cactus Plain (AZGFD)	Hoodoo Wash (AZGFD)	Palomas Plain (AZGFD)
SITE CHARACTERISTICS					
1. Total area of site (acres)	14,500	4,000	11,000	14,000	25,000
BLM acres	14,500	4,000	11,000	13,715	23,440
Private acres				3	
State Trust acres				282	1,560
2. Sources of data for the site.	AGFD GIS and SWAP data	AGFD GIS and SWAP data	AGFD GIS and SWAP data	AGFD GIS and SWAP data	AGFD GIS and SWAP data
3. Mitigates for all or most identified residual impacts that may warrant compensatory mitigation. √ for yes (1 point) or X for no (–2 points); Include justification.	√ Ecological Resources: Vegetation, Wildlife, SSS Animals; Visual Resources	√ Ecological Resources: Vegetation, Wildlife, SSS Animals; Visual Resources	√ Ecological Resources: SSS Animals, Wildlife, Vegetation; Visual Resources	√ Ecological Resources: SSS Animals, Wildlife, Vegetation; Visual Resources	√ Ecological Resources: SSS Animals, Wildlife, Vegetation; Visual Resources
4. Mitigation action (restoration/enhancement, acquisition, withdrawal, special designation, etc.).	Habitat Enhancement Restoration, Species Specific Management Action Visual Resources	Habitat Enhancement Restoration, Species Specific Management Action Visual Resources	Habitat Enhancement Riparian System Rehabilitation or Restoration Species Specific Management Action Visual Resources	Habitat Enhancement Riparian System Rehabilitation or Restoration Species Specific Management Action Visual Resources	Habitat Enhancement Riparian System Rehabilitation or Restoration Species Specific Management Action Visual Resources
5. Site and its proposed actions meet regional conservation/ mitigation goals and objectives. √ for yes (1 point) or X for no (–2 points).	√	√	√	√	√
Justification.	Site has creosote bursage, Sonoran Paloverde-Mixed Cacti Desert Scrub and 24 vertebrate special status species	Site has creosote bursage, Sonoran Paloverde-Mixed Cacti Desert Scrub and 23 vertebrate special status species	Site has creosote bursage, Sonoran Paloverde-Mixed Cacti Desert Scrub and 24 vertebrate special status species	Site has creosote bursage, Sonoran Paloverde-Mixed Cacti Desert Scrub and 28 vertebrate special status species	Site has creosote bursage, Sonoran Paloverde-Mixed Cacti Desert Scrub and 27 vertebrate special status species
6. Proposed Mitigation Action and location Consistent with the Resource Management Plan. √ for yes (1 point) or X for no (–2 points); Include justification.	√ Site is in Yuma RMP Palomas Plain and Desert Mountains Wildlife Habitat Management Areas	√ Site is in Yuma RMP Palomas Plain Wildlife Habitat Management Area	√ Site is in Havasu RMP Wildlife Habitat Management Area	√ Site is in Yuma RMP Palomas Plain and Desert Mountains Wildlife Habitat Management Areas	√ Site is in Yuma RMP Palomas Plain and Desert Mountains Wildlife Habitat Management Areas
7. Same HUC 4 watershed. Specify watershed.	Lower Colorado	Lower Gila	Lower Colorado	Lower Gila	Lower Gila
8. VRI Class and acres associated with each class.	Class III: 14,561 acres	Class II: 2,556 acres; Class III: 1,607 acres	Class II: 12,173 acres; Class III: 37,788; Class IV: 8,897	Class III: 11,202 acres; Class III: 3,1220 acres	Class II: 24,951 acres; Class III: 17 acres

Regional Mitigation Strategy for the Arizona SEZs

Criteria	Candidate Sites				
	Ranegras Plain (AZGFD)	Sacaton Flats (AZGFD)	Cactus Plain (AZGFD)	Hoodoo Wash (AZGFD)	Palomas Plain (AZGFD)
9. Similar landscape value, ecological functionality, biological value, species, habitat types, and/or natural features. Score based on responses to criteria 9a and 9b. √ for yes (1 point) or X for no (-2 points).	√	√	√	√	● √
9a. Current terrestrial landscape intactness score (use Sonoran Desert Rapid Ecological Assessment Data) and acres associated with each intactness category.[33]	Very Low: 102 acres; Low: 1,816 acres; Mod Low: 1,729 acres; Mod High: 1,916 acres; High: 1,916 acres; Very High: 7,098 acres	Very Low: 10 acres; Low: 233 acres; Mod Low: 520 acres; Mod High: 1,414 acres; High: 218 acres; Very High: 1,768 acres	Mod Low: 2,2825 acres; Mod High: 7,689 acres; High: 1,008 acres; Very High: 121 acres	Very Low: 104 acres; Low: 38 acres; Mod Low: 608 acres; Mod High: 3,664 acres; High: 6,291 acres; Very High: 3,903 acres	Low: 325 acres; Mod Low: 3,962 acres; Mod High: 8,295 acres; High: 4,842 acres; Very High: 9,099 acres
9b. Dominant vegetation communities.	Sonoran-Mojave Creosotebush-White Bursage Desert Scrub (99%) Sonoran Paloverde-Mixed Cacti Desert Scrub (41%)	Sonoran-Mojave Creosotebush-White Bursage Desert Scrub (89%) Sonoran Paloverde-Mixed Cacti Desert Scrub (11%)	Sonoran-Mojave Creosotebush-White Bursage Desert Scrub (67%) Sonoran Paloverde-Mixed Cacti Desert Scrub (32%)	Sonoran-Mojave Creosotebush-White Bursage Desert Scrub (98%) Sonoran Paloverde-Mixed Cacti Desert Scrub (2%)	Sonoran-Mojave Creosotebush-White Bursage Desert Scrub (96%) Sonoran Paloverde-Mixed Cacti Desert Scrub (4%)
10. In SEZ Ecoregion √ for yes (1 point) or X for no (-2 points).	√	√	√	√	√
11. In SEZ ecological subregion. Specify subregion. √ for yes (1 point) or X for no (-2 points).	√	√	√		√
12. Provides adequate geographic extent. Depending on whether site provides area for mitigation at least as large as the entire developable area of the SEZ. √ for yes (1 point) or X for no (-2 points).	√	√	√	√	√
FEASIBILITY					
13. Feasibility of action	3	3	3	3	3
Justification of feasibility score. Scores for 13a through 13 e were provided by stakeholders. BLM used these scores as well as their knowledge of the sites and actions.	Site has good access, no major land use restrictions, washes and areas for habitat enhancement restoration.	Site has good access, no major land use restrictions, washes and areas for habitat enhancement restoration.	Site has good access, no major land use restrictions, washes and areas for habitat enhancement restoration.	Site has good access, no major land use restrictions, washes and areas for habitat enhancement restoration.	Site has good access, no major land use restrictions, washes and areas for habitat enhancement restoration.
13a. What level of documentation is available to demonstrate effectiveness of mitigation action? Use scale of 1 (little to no documentation) to 5 (well-documented).	3	3	3	3	3

---

33 Intactness Categories: Very Low (-1.0 – -0.75); Low (-0.75 – -0.5); Mod Low (-0.5 – 0.0); Mod High (0.0 – 0.5); High (0.5 – 0.75) Very High (0.75 – 1.0).

Regional Mitigation Strategy for the Arizona SEZs

Criteria	Candidate Sites				
	Ranegras Plain (AZGFD)	Sacaton Flats (AZGFD)	Cactus Plain (AZGFD)	Hoodoo Wash (AZGFD)	Palomas Plain (AZGFD)
Justification.	Techniques well documented, nothing specifically on site.	Techniques are well documented, nothing specifically on site.	Techniques are well documented, nothing specifically on site.	Techniques are well documented, nothing specifically on site.	Techniques are well documented, nothing specifically on site.
13b.[34] Based on action required (e.g., restoration, BLM land management action, land acquisition, Congressional action), how difficult will implementation be? Use scale of 1 (difficult) to 5 (relatively easy).	4	4	4	4	4
Justification.	Site has motorized access.	Site has motorized access.	Candidate site is accessible by motorized vehicle making conducting actions easier.	Candidate site is accessible by motorized vehicle making conducting actions easier.	Candidate site is accessible by motorized vehicle making conducting actions easier.
13c. Time frame needed to establish site as mitigation location (estimated years).	1–2 years	1–2 years	1–2 years	1–2 years	1–2 years
13d. Time frame for achieving mitigation goals and objectives from implementation (estimated years).	5 years	5 years	5 years	5 years	5 years
13e. Cost estimate.	$48,000	$48,000	$48,000	$48,000	$48,000
EFFECTIVENESS / ADDITIONALITY					
14. Effectiveness and Additionality	3	3	3	3	3
Justification of effectiveness and additionality. Scores for 14a through 14c were provided by stakeholders. BLM used these scores as well as their knowledge of the sites and actions.	Narrow focus on riparian restoration/creation	Narrow focus on riparian restoration/creation	Narrow focus on riparian restoration/creation	Narrow focus on riparian restoration/creation	Narrow focus on riparian restoration/creation
14a.[35] To what extent can the full spectrum of mitigation goals/objectives be met simultaneously? Use scale of 0 (low) to 5 (high).	4	4	3	4	4

---

[34] Rate the mitigation action, based on the following scale: restoration/enhancement actions (score of 5); BLM planning decisions (score of 3–4); land acquisition actions (score of 1–3); Congressional actions (score of 1). Ratings should be adjusted on the basis of factors such as cost of the action; time and effort requirements; public and/or BLM support for or opposition to action; and, for land acquisitions, willingness of seller.

[35] Rate the extent to which the mitigation desired outcomes can be met simultaneously through mitigation actions at the site, based on the following scale: all (100%) of the goals and objectives can be met (score of 5); 75–99% can be met (score of 4); 50–74% (score of 3); 25–49% can be met (score of 2); less than 25% can be met (score of 1); none of the goals/objectives can be met (score of 0).

Regional Mitigation Strategy for the Arizona SEZs

Criteria	Candidate Sites				
	Ranegras Plain (AZGFD)	Sacaton Flats (AZGFD)	Cactus Plain (AZGFD)	Hoodoo Wash (AZGFD)	Palomas Plain (AZGFD)
Justification.	Meet objectives for biological resources, VRM Class 3, cultural unknown	Meet objectives for biological resources, VRM Classes 2 AND 3, cultural unknown	Meet objectives for biological resources, VRM 3, cultural unknown	Meet objectives for biological resources, VRM 2, cultural unknown	Meet objectives for biological resources, VRM 2, cultural unknown
14b. [36] How effective will the mitigation be in the context of achieving mitigation goals/objectives for conserving/restoring ecosystem intactness? Use scale of 1 (low) to 5 (high).	5	4	4	5	5
Justification.	Area undisturbed by human infrastructure, some grazing impacts.	Area undisturbed by human infrastructure, some grazing impacts, site is smaller.	Area undisturbed by human infrastructure, nearby agriculture may impact.	Area undisturbed by human infrastructure, some grazing impacts.	Area undisturbed by human infrastructure, some grazing impacts.
14c. Mitigation consists of actions that would not otherwise be undertaken by BLM.	✓	✓	✓	✓	✓
RISK					
15. Risk of action(s)	3	3	3	3	3
Justification of risk score. Scores for 15a through 15b were provided by stakeholders. BLM used these scores as well as their knowledge of the sites and actions.					
15a. What are the constraints or threats to success?	Conflicts with recreation and grazing.	Conflicts with recreation.	Conflicts with recreation and nearby agriculture.	Conflicts with recreation and nearby agriculture.	Conflicts with recreation.
15b. What are surrounding land uses that will impact mitigation success (e.g., proximity to expanding urban areas, pressures on region for recreational land use, excessive groundwater withdrawal and drawdown conditions that could affect resources on the mitigation site)?	4 Some recreation impacts	4 Some recreation impacts	Nearby agriculture activity may result in groundwater draw down	Some recreation/grazing impacts	Some recreation/grazing impacts
DURABILITY					
16. Durability of action(s)	3	3	3	3	3

---

[36] Rate the effectiveness of the mitigation actions at the site in terms of achieving mitigation goals/objectives, based on the following scale: highly effective (score of 5); moderately effective (scores of 2–4), and minimally effective (score of 1).

Regional Mitigation Strategy for the Arizona SEZs

Criteria	Candidate Sites				
	Ranegras Plain (AZGFD)	Sacaton Flats (AZGFD)	Cactus Plain (AZGFD)	Hoodoo Wash (AZGFD)	Palomas Plain (AZGFD)
Justification of durability score. Scores for 16a through 16b were provided by stakeholders. BLM used these scores as well as their knowledge of the sites and actions.					
16a.[37] How durable would the mitigation be from a time frame and management perspective? Use scale of 1 (low) to 5 (high).	3	3	3	3	3
Justification.	BLM Land Identified in RMP as a Wildlife Habitat Area	BLM Land Identified in RMP as a Wildlife Habitat Area	BLM Land Identified in RMP as a Wildlife Habitat Area	BLM Land Identified in RMP as a Wildlife Habitat Area	BLM Land Identified in RMP as a Wildlife Habitat Area
16b. Are there potential effects of future climate change[38]?	High	Moderate	Moderate	Moderate	Moderate
PRELIMINARY SCORING Calculate score by summing the entries in blue-shaded cells. Scores are calculated based on entries in blue-shaded cells as follows: all scaled values (i.e., ratings from 1 to 5) are summed; 1 point is added for each √; 2 points are deleted for each X.	19	19	19	19	19
ADDITIONAL CONSIDERATIONS					
17. Presence of unique/valuable resources or features. (Up to 3 additional points for unique/valuable resources or features present at the candidate site, in 17a through 17h.)	1	1	1	1	1
17a. Perennial, protected sources of water	Desert washes	Desert washes	Desert washes	Desert washes	North American Warm Desert Wash

---

[37] Rate the temporal and managerial durability of the mitigation action, based on the following scale: Congressionally protected lands would be very durable (score of 5); other federally administered lands specifically designated in land use plans or withdrawn by public land order would be moderately to very durable (score of 4–5); federally administered lands without any special designation but with enforcement oversight would have limited durability (score of 2); lands without special designation or enforcement oversight would not be very durable (score of 1).

[38] Climate change categories are from the Sonoran Desert Rapid Ecoregional Assessment future climate change model (BLM 2011).

Regional Mitigation Strategy for the Arizona SEZs

Criteria	Candidate Sites				
	Ranegras Plain (AZGFD)	Sacaton Flats (AZGFD)	Cactus Plain (AZGFD)	Hoodoo Wash (AZGFD)	Palomas Plain (AZGFD)
17b. Unique species assemblages	Bald Eagle, Gila Monster, Sonoran Desert Tortoise, American Bittern, American Beaver, Arizona Bell's Vireo, Arizona Pocket Mouse, California Leaf-nosed Bat, Cave Myotis, Gila Woodpecker, Gilded Flicker, Greater Western Mastiff Bat, Harquahala Southern Pocket Gopher, Harris' Antelope Squirrel, Kit fox, Le Conte's Thrasher, Lincoln's Sparrow, Little Pocket Mouse, Mexican Free-tailed Bat, Pacific Wren, Pale Townsend's Big-eared bat, Pocketed Free-Tailed Bat, Sonoran Desert Toad, Spotted Bat, Western Burrowing Owl, Western Red Bat, Wood Duck, Yuma Myotis	Gila Monster, Sonoran Desert Tortoise, Arizona Pocket Mouse, California Leaf-nosed Bat, Cave Myotis, Desert Bighorn Sheep, Gila Woodpecker, Gilded Flicker, Greater Western Mastiff Bat, Harris' Antelope Squirrel, Kit fox, Le Conte's Thrasher, Lincoln's Sparrow, Little Pocket Mouse, Mexican Free-tailed Bat, Pale Townsend's Big-eared bat, Pocketed Free-Tailed Bat, Sonoran Desert Toad, Spotted Bat, Yuma Myotis	American Peregrine Falcon, Bald Eagle, Gila Monster, Lowland Leopard Frog, Sonoran Desert Tortoise, Arizona Bell's Vireo, Arizona Pocket Mouse, California Leaf-nosed Bat, Cave Myotis, Gila Woodpecker, Gilded Flicker, Greater Western Mastiff Bat, Harris' Antelope Squirrel, Kit fox, Le Conte's Thrasher, Little Pocket Mouse, Mexican Free-tailed Bat, Pale Townsend's Big-eared bat, Pocketed Free-tailed Bat, Regal Horned Lizard, Sonoran Desert Toad, Spotted Bat, Yuma Myotis	Bald Eagle, Gila Monster, Sonoran Desert Tortoise, American Beaver, American Bittern, Arizona Bell's Vireo, Arizona Pocket Mouse, California Leaf-nosed Bat, Cave Myotis, Desert Bighorn Sheep, Ferruginous Hawk, Gila Woodpecker, Gilded Flicker, Greater Western Mastiff Bat, Harquahala Southern Pocket Gopher, Harris' Antelope Squirrel, Kit fox, Le Conte's Thrasher, Lincoln's Sparrow, Little Pocket Mouse, Pacific Wren, Pale Townsend's Big-eared bat, Pocketed Free-tailed Bat, Savannah Sparrow, Sonoran Desert Toad, Wood Duck, Yuma Myotis	Bald Eagle, Gila Monster, Sonoran Desert Tortoise, American Bittern, American Beaver, Arizona Bell's Vireo, Arizona Pocket Mouse, California Leaf-nosed Bat, Cave Myotis, Desert Bighorn Sheep, Gila Woodpecker, Gilded Flicker, Golden Eagle, Greater Western Mastiff Bat, Harris' Antelope Squirrel, Harquahala Southern Pocket Gopher, Kit fox, Le Conte's Thrasher, Lincoln's Sparrow, Little Pocket Mouse, Le Conte's Thrasher, Lincoln's Sparrow, Little Pocket Mouse, Mexican Free-tailed Bat, Pale Townsend's Big-eared bat, Pocketed Free-Tailed Bat, Sonoran Desert Bat, Wood Duck, Yuma Myotis
17c. AZGFD Species of Greatest Conservation Need (distribution models)/Heritage Data Management System species (occurrence data)	24/8	23/4	24/7	28/7	27/6
17d. BLM categorized desert tortoise habitat	East side touches Category 3	North half of site in Category 2	Category 2 and 3 on edge	Category 3 on edge	Category 2 on edge
17e. T&E species or critical habitat and/or BLM sensitive species	T&E Species: Sonoran Pronghorn Nonessential Experimental Population; Critical Habitat: none	T&E Species: Sonoran Pronghorn Nonessential Experimental Population; Critical Habitat: none	No T&E species or critical habitat	T&E Species: Sonoran Pronghorn Nonessential Experimental Population; Critical Habitat: none	T&E Species: Sonoran Pronghorn Nonessential Experimental Population; Critical Habitat: none
17f. Desert washes (miles) or ephemeral playas (acres)	Desert washes	Desert washes	Desert washes	Desert washes	North American Warm Desert Wash
17g. Known highly significant and unique cultural resources					
17h. Other	Xeric riparian			Xeric riparian, riparian	Riparian, xeric riparian

Regional Mitigation Strategy for the Arizona SEZs

Criteria	Candidate Sites					
	Ranegras Plain (AZGFD)	Sacaton Flats (AZGFD)	Cactus Plain (AZGFD)	Hoodoo Wash (AZGFD)	Palomas Plain (AZGFD)	
18. Links two or more protected areas. √ for yes (1 point) or 0 for no (no score adjustment); include justification.			0	0	0	
COMBINED SCORE Add preliminary score to the additional consideration criteria in the blue-shaded cells. Scores are calculated based on entries in blue-shaded cells as follows: scaled values (i.e., ratings from 1 to 3) are summed; 1 point is added for each √	20	20	20	20	20	

Regional Mitigation Strategy for the Arizona SEZs

	Candidate Sites				
Criteria	Sears Point ACEC (TNC, BLM Yuma FO and Archaeology Southwest)	Saddle Mountain ACEC (TWS)	Fred J. Weiler Vegetation Habitat Management Area (BLM Yuma FO)	Quail Point (Archaeology Southwest)	Lower Gila River Terraces Package (Archaeology Southwest)
SITE CHARACTERISTICS					
1. Total area of site (acres)	28,436	5,500 acres and another 12,040 acres of surrounding desert flats and foothills were determined to possess wilderness characteristics.	Surrounds and overlaps approximately 12,400 of the Sears Point ACEC	360	1,641[39]
BLM acres	13,644	17,540			1,286
Private acres	8,638			360	302
State Trust acres	6,154				
2. Sources of data for the site.	USGS Protected Areas Database of the United States, TNC Ecoregional rollup, BLM REA, Yuma RMP, STATSGO, SURGO for soil stability; wildlife linkages.	BLM Lower Sonoran RMP	Yuma RMP	Archeology Southwest	Archaeology Southwest
3. Mitigates for all or most identified residual impacts that may warrant compensatory mitigation. V for yes (1 point) or X for no (-2 points); Include Justification.	V Ecological Resources: SSS Animals, Wildlife, Vegetation; Cultural Resources	V Ecological Resources: Vegetation, Wildlife; Cultural resources	V Ecological Resources: Vegetation, Wildlife, SSS Animals; Cultural Resources	V Cultural Resources	V Cultural Resources
4. Mitigation action (restoration/enhancement, acquisition, withdrawal, special designation, etc.).	Acquisition; restoration; withdrawal of an additional 4,900 acres of federal land within the ACEC; develop a Sears Point ACEC plan.	Restoration of closed roads; Off Highway Vehicle law enforcement; purchase and protection of private lands in the adjacent wildlife movement corridor (there is not a contiguous strip of BLM land in the corridor); RMP amendments for ROW exclusion or other protection.	Pursue the withdrawal of an additional 4,900 acres of federal land within the ACEC. Develop a Sears Point ACEC plan. Seek to acquire non-federal lands and interests within or adjacent to lands within the ACEC.	Land Acquisition	Land Acquisition and On-Site Restoration
5. Site and its proposed actions meet regional conservation/mitigation goals and objectives. V for yes (1 point) or X for no (-2 points).	V	V	V	V	V

---

[39] Total acreage of combined sites: Red Rock Canyon 1,189 acres (1,286 BLM acres and 302 private acres), Robbins Butte 97 BLM acres, Gillespie Dam North 89 private acres, Gillespie Dam South 151 private acres, Enterprise North 46 private acres, Enterprise Central 16 private acres, Enterprise South 53 private acres.

Regional Mitigation Strategy for the Arizona SEZs

Criteria	Candidate Sites				
	Sears Point ACEC (TNC, BLM Yuma FO and Archaeology Southwest)	Saddle Mountain ACEC (TWS)	Fred J. Weiler Vegetation Habitat Management Area (BLM Yuma FO)	Quail Point (Archaeology Southwest)	Lower Gila River Terraces Package (Archaeology Southwest)
Justification.	Preserves and restores creosote/bursage habitat, preserves and restores riparian habitat, protects BLM sensitive species, protects cultural resources. The Sears Point ACEC was designated as such because of the archaeological district as well as the prominent basalt mesas, historic trail corridors and important riparian vegetation including a mesquite bosque and the Fred J. Weiler Greenbelt.	BLM recognized the values in the region of this mitigation site in the Lower Sonoran RMP through the LWC management, the designation of the Saddle Mountain Outstanding Natural Area ACEC, and designation of the Gila Bend Mtns to Saddle Mtn wildlife movement corridor. Strengthening protective management for these areas and expanding onto adjacent land with similar resources and values will help meet conservation/mitigation goals and objectives.	Identified for acquisition and restoration in Yuma and Lower Sonoran RMPs	Within Sears Point ACEC expansion area (Yuma RMP)	Protection of cultural resources and restoration of habitat.
6. Proposed Mitigation Action and location Consistent with the Resource Management Plan. √ for yes (1 point) or X for no (-2 points); Include justification.	√	√	√	√	√
7. Same HUC 4 watershed. Specify watershed.	Lower Gila	Lower Gila	Lower Gila	Lower Gila	Lower Gila
8. VRI Class and acres associated with each class.	Class II: 11,903 acres; Class III: 1309 acres; Class IV: 3 acres	Class II: 13,027 acres; Class III: 4,896 acres; Class IV: 30,587 acres	Class II: 12,703 acres; Class III: 1,353 acres; Class IV: 4 acres	No VRI	Class II: 59 acres; Class III: 535 acres; Class IV: 696 acres
9. Similar landscape value, ecological functionality, biological value, species, habitat types, and/or natural features. Score based on responses to criteria 9a and 9b. √ for yes (1 point) or X for no (-2 points).	√	√	√	√	√

Regional Mitigation Strategy for the Arizona SEZs

		Candidate Sites			
Criteria	Sears Point ACEC (TNC, BLM Yuma FO and Archaeology Southwest)	Saddle Mountain ACEC (TWS)	Fred J. Weiler Vegetation Habitat Management Area (BLM Yuma FO)	Quail Point (Archaeology Southwest)	Lower Gila River Terraces Package (Archaeology Southwest)
9a. Current terrestrial landscape intactness score (use Sonoran Desert Rapid Ecological Assessment Data) and acres associated with each intactness category.[40]	Very Low: 8,566 acres; Low: 10,359 acres; Mod Low: 7,948 acres; Mod High: 1,237 acres; High: 317 acres	Very Low: 4,430 acres; Low: 5,934 acres; Mod Low: 8,663 acres; Mod High: 15,631 acres; High: 14,722 acres; Very High: 6,178	Very Low: 4,087 acres; Low: 7,090 acres; Mod Low: 2,962 acres; Mod High: 321 acres;	Very Low: 12 acres; Low: 199 acres; Mod Low: 170 acres	Very Low: 270 acres; Low: 305 acres; Mod Low: 346 acres; Mod High: 418 acres; High: 2 acres; Very High: 300 acres
9b. Dominant vegetation communities.	Sonoran-Mojave Creosotebush-White Bursage Desert Scrub (70%) Introduced Vegetation (14%)	Sonoran-Mojave Creosotebush-White Bursage Desert Scrub (68%) Sonoran Paloverde-Mixed Cacti Desert Scrub (31%)	Sonoran-Mojave Creosotebush-White Bursage Desert Scrub (54%) Introduced Vegetation (26%)	Sonoran-Mojave Creosotebush-White Bursage Desert Scrub (89%) Sonoran Paloverde-Mixed Cacti Desert Scrub (10%)	Sonoran-Mojave Creosotebush-White Bursage Desert Scrub (88%) Sonoran Paloverde-Mixed Cacti Desert Scrub (9%)
10. In SEZ Ecoregion √ for yes (1 point) or X for no (-2 points).	√	√	√	√	√
11. In SEZ ecological subregion. Specify subregion. √ for yes (1 point) or X for no (-2 points).	√ Colorado Desert - EPA: 81j, 81g, 81m	√ 81j Central Sonoran/Colorado Desert Basins	√	√	√
12. Provides adequate geographic extent? Depending on whether site provides area for mitigation at least as large as the entire developable area of the SEZ. √ for yes (1 point) or X for no (-2 points).	√	√	√	X	√
FEASIBILITY					
13. Feasibility of action	5	3	5	5	4
Justification of action feasibility. Stakeholder-provided score for 13a through 13e, as well as BLM expert knowledge, were used by the BLM to arrive at the overall Feasibility score, which is included in the overall site score.	Closure and revegetation of roads is straightforward and low cost. Restoration of agriculture fields can range from passive and cheap to very active and moderately expensive, depending on goals and resources available. Land acquisition can be complex, but is not critical to success at this site.	Mitigation and landscape restoration for incompatible activities identified in RMP. Retention and acquisition of lands in ACEC identified in RMP. RMP amendment for expansion of ACEC, however, adds time and complexity.	Restoration identified in RMP	Acquisition and restoration identified in RMP	Acquisition and restoration identified in RMP

[40] Intactness Categories: Very Low (-1.0 – -0.75); Low (-0.75 – -0.5); Mod Low (-0.5 – 0.0); Mod High (0.0 – 0.5); High (0.5 – 0.75) Very High (0.75 – 1.0).

Regional Mitigation Strategy for the Arizona SEZs

Criteria	Candidate Sites				
	Sears Point ACEC (TNC, BLM Yuma FO and Archaeology Southwest)	Saddle Mountain ACEC (TWS)	Fred J. Weiler Vegetation Habitat Management Area (BLM Yuma FO)	Quail Point (Archaeology Southwest)	Lower Gila River Terraces Package (Archaeology Southwest)
13a. What level of documentation is available to demonstrate effectiveness of mitigation action? Use scale of 1 (little to no documentation) to 5 (well-documented).	4	4	4	5	5
Justification.	Road closure and revegetation is a widespread practice with good results. Agricultural field revegetation is a widespread practice with mixed results.	BLM has a long history of using protective management to meet goals and objectives for resources and to support multiple use and sustained yield of the varied resources and values found on public lands. This includes designations such as ACECs and establishment of protective management prescriptions through RMPs.			
13b. [41] Based on action required (e.g., restoration, BLM land management action, land acquisition, Congressional action), how difficult will implementation be? Use scale of 1 (difficult) to 5 (relatively easy).	4	3	5	5	
Justification.	Proposed mitigation actions already identified in Yuma RMP, no RMP amendment needed to undertake them.	RMP is recent, and revision not anticipated soon. Directed RMP amendment could occur, subject to funding and/or need.	Actions identified in RMP	Actions identified in RMP	Actions identified in RMP
13c. Time frame needed to establish site as mitigation location (estimated years).	2 years	1–2 years	2 years	2 years	2 years if acquisition
13d. Time frame for achieving mitigation goals and objectives from implementation (estimated years).	5	1–4 years depending on actions	5–10 years	0 years—goals achieving primarily with acquisition	2–5 years

[41] Rate the mitigation action based on the following scale: restoration/enhancement actions (score of 5); BLM planning decisions (score of 5); land acquisition actions (score of 3–4); Congressional actions (score of 1). Ratings should be adjusted on the basis of factors such as cost of the action; time and effort requirements; public and/or BLM support for or opposition to action; and, for land acquisitions, willingness of seller.

Regional Mitigation Strategy for the Arizona SEZs

Criteria	Candidate Sites				
	Sears Point ACEC (TNC, BLM Yuma FO and Archaeology Southwest)	Saddle Mountain ACEC (TWS)	Fred J. Weiler Vegetation Habitat Management Area (BLM Yuma FO)	Quail Point (Archaeology Southwest)	Lower Gila River Terraces Package (Archaeology Southwest)
13e. Cost estimate.	$35,000 to $18,600,000	$100,000 to $500,000	$34,534,000[42]	$100,000 to $300,000	$150,000 to $300,000 for acquisition
EFFECTIVENESS / ADDITIONALITY					
14. Effectiveness and Additionality	5	4	4	5	5
Justification of effectiveness and additionality score. Scores for 14a through 14c were provided by stakeholders. BLM used these scores as well as their knowledge of the sites and actions.	Wide range of opportunities	Acquisition increases additionality score	Restoration identified in RMP. Not as diverse as far as opportunities.	Acquisition and restoration identified in RMP	Acquisition and restoration identified in RMP. Collectively, as a unit, the small parcels offer several additional opportunities.
14a.[43] To what extent can the full spectrum of mitigation goals/objectives be met simultaneously? Use scale of 0 (low) to 5 (high).	5	4	4	4	
Justification.	The mitigation site includes a wide range of resources and values that would benefit from additional protections, including most of the resources and values that would be impacted from development in the SEZ.				
14b.[44] How effective will the mitigation be in the context of achieving mitigation goals/objectives for conserving/restoring ecosystem intactness? Use scale of 1 (low) to 5 (high).	5	4	4	4	

---

[42] Based on clearing and tall pot revegetation of best 1,240 acres at $25,850 per acre.

[43] Rate the extent to which the mitigation desired outcomes can be met simultaneously through mitigation actions at the site, based on the following scale: all (100%) of the goals and objectives can be met (score of 5); 75–99% can be met (score of 4); 50–74% (score of 3); 25–49% can be met (score of 2); less than 25% can be met (score of 1); none of the goals/objectives can be met (score of 0).

[44] Rate the effectiveness of the mitigation actions at the site in terms of achieving mitigation goals/objectives, based on the following scale: highly effective (score of 5); moderately effective (scores of 2–4), and minimally effective (score of 1).

Regional Mitigation Strategy for the Arizona SEZs

Criteria	Candidate Sites				
	Sears Point ACEC (TNC, BLM Yuma FO and Archaeology Southwest)	Saddle Mountain ACEC (TWS)	Fred J. Weiler Vegetation Habitat Management Area (BLM Yuma FO)	Quail Point (Archaeology Southwest)	Lower Gila River Terraces Package (Archaeology Southwest)
Justification.	Acquisition and restoration preserve and increase intactness and connectivity of landscapes and habitat.	By connecting several existing protected areas and increasing protections for a BLM-designated wildlife movement corridor, the mitigation site would be very effective in achieving goals/objectives for conserving/restoring ecosystem intactness.			
14c. Mitigation consists of actions that would not otherwise be undertaken by BLM	Funding through LWCF or other means has not been available for acquisition/restoration of Sears Point.	Funds needed to protect mitigation sites and restoration and management of those sites.	∨	∨	∨
RISK					
15. Risk of action(s)	4	4	4	5	5
Justification of risk score. Scores for 15a through 15b were provided by stakeholders. BLM used these scores as well as their knowledge of the sites and actions.					
15a. What are the constraints or threats to success?		Possible public opposition to RMP amendment			
15b. What are surrounding land uses that will impact mitigation success (e.g., proximity to expanding urban areas, pressures on region for recreational land use, excessive groundwater withdrawal and drawdown conditions that could affect resources on the mitigation site)?	Limited agriculture and other development on inholdings within ACEC expansion area. Private development adjacent to south side, agriculture, and groundwater pumping within site.	Low-density residential, industrial, and I-10 corridor	Agricultural development and groundwater pumping	Some recreational use	Agricultural development
DURABILITY					
16. Durability of action(s)	4	4	4	4	4
Justification of durability score. Scores for 16a and 16b were provided by stakeholders. BLM used these scores as well as their knowledge of the sites and actions.		Actions within ACEC are more durable than actions outside of the ACEC.			

Regional Mitigation Strategy for the Arizona SEZs

Criteria	Candidate Sites				
	Sears Point ACEC (TNC, BLM Yuma FO and Archaeology Southwest)	Saddle Mountain ACEC (TWS)	Fred J. Weiler Vegetation Habitat Management Area (BLM Yuma FO)	Quail Point (Archaeology Southwest)	Lower Gila River Terraces Package (Archaeology Southwest)
16a. [45] How durable would the mitigation be from a time frame and management perspective? Use scale of 1 (low) to 5 (high).	4	4	4	4	4
Justification.		BLM has a variety of special designations and management actions at its disposal to establish the necessary level of durability to fulfill mitigation desired outcomes. BLM can add durability by creating overlapping protective designations and committing that if a mitigation site were to lose protective management, the agency would protect another, equivalent site to maintain an equal level of mitigation. Mitigation funds would provide a durable source of funds for management.			
16b. Are there potential effects of future climate change [46]?	Moderate	High	Moderate	High	
PRELIMINARY SCORING Calculate score by summing the entries in blue-shaded cells. Scores are calculated based on entries in blue-shaded cells as follows: all scaled values (i.e., ratings from 1 to 5) are summed; 1 point is added for each √; 2 points are deleted for each X.	25	22	24	23	25

[45] Rate the temporal and managerial durability of the mitigation action, based on the following scale: Congressionally protected lands would be very durable (score of 5); other federally administered lands specifically designated in land use plans or withdrawn by public land order would be moderately to very durable (score of 4–5); federally administered lands without any special designation but with enforcement oversight would have limited durability (score of 2); lands without special designation or enforcement oversight would not be very durable (score of 1).

[46] Climate change categories are from the Sonoran Desert Rapid Ecoregional Assessment future climate change model (BLM 2011).

Regional Mitigation Strategy for the Arizona SEZs

Criteria	Candidate Sites				
	Sears Point ACEC (TNC, BLM Yuma FO and Archaeology Southwest)	Saddle Mountain ACEC (TWS)	Fred J. Weiler Vegetation Habitat Management Area (BLM Yuma FO)	Quail Point (Archaeology Southwest)	Lower Gila River Terraces Package (Archaeology Southwest)
ADDITIONAL CONSIDERATIONS					
17. Presence of unique/valuable resources or features. (Up to 3 additional points for unique/valuable resources or features present at the candidate site, in 17a through 17h.)	3	2	1	3	2
17a. Perennial, protected sources of water					
17b. Unique species assemblages					
17c. AZGFD Species of Greatest Conservation Need (distribution models)/ Heritage Data Management System species (occurrence data)	36/2		38/-		
17d. BLM categorized desert tortoise habitat		Category 2			Category 2 (Red Rock Canyon only)
17e. T&E species or critical habitat and/or BLM sensitive species	No T&E species or critical habitat. BLM sensitive species: Lowland Leopard Frog, Western Burrowing Owl, Gilded Flicker, Bald Eagle, Le Conte's Thrasher, Pale Townsend's big-eared bat, Spotted Bat, Greater Western Mastiff Bat, CA leaf-nosed bat, Cave Myotis	T&E Species: Sonoran Pronghorn Nonessential Experimental Population; Critical Habitat: none	No T&E species or critical habitat	No T&E species or critical habitat	Proposed critical habitat for Western yellow-billed cuckoo (Robbins Butte only)
17f. Desert washes (miles) or ephemeral playas (acres)	Desert washes (4.7 miles)				
17g. Known highly significant and unique cultural resources	High petroglyph concentration, National Historic Trail. The ACEC contains a 3,700-acre core area that includes a high concentration of petroglyphs which are within an NRHP-listed archaeological district.	Petroglyphs, rock shelter, geoglyphs		Rock art, precontact trail, rock shrines, and geoglyph	Petroglyphs, Archaic, Hohokam, and Patayan archaeology (including habitations with ballcourt features)
17h. Other	Gila River riparian area (15.5 miles) Sacred areas, large mesquite bosques	Unique geological formation	Large mesquite bosques	Sacred areas	

Regional Mitigation Strategy for the Arizona SEZs

Criteria	Candidate Sites				
	Sears Point ACEC (TNC, BLM Yuma FO and Archaeology Southwest)	Saddle Mountain ACEC (TWS)	Fred J. Weiler Vegetation Habitat Management Area (BLM Yuma FO)	Quail Point (Archaeology Southwest)	Lower Gila River Terraces Package (Archaeology Southwest)
18. Links two or more protected areas √ for yes (1 point) or 0 for no (no score adjustment); Include justification.	0	√ Acquisition of private lands between Saddle and Gila Bend Mountains	0	0	0
COMBINED SCORE Add preliminary score to the additional consideration criteria in the blue-shaded cells. Scores are calculated based on entries in blue-shaded cells as follows: scaled values (i.e., ratings from 1 to 3) are summed; 1 point is added for each √.	28	25	25	26	27

Regional Mitigation Strategy for the Arizona SEZs

Criteria		Candidate Sites					
		La Osa Ranch (Pima County)	Marana Mound (Pima County)	Ajo (Pima County)	Boa Sorte (Pima County)	Cocoraque Butte (Pima County)	Los Robles Archeological/Historical District (Pima County)
SITE CHARACTERISTICS							
1. Total area of site (acres)		6,000 to 12,000	13,000 acres	500 acres	1,800 acres	4,628 acres	3,136 acres
	BLM acres						
	Private acres	6,000 to 12,000	540 acres	500 acres	1,800 acres	4,628 acres	3,136 acres
	State Trust acres		12,460 acres				
2. Sources of data for the site.		Arizona HabiMap, Land Advisors Organization		Arizona HabiMap, Sonoran Desert Conservation Plan	Arizona HabiMap, Sonoran Desert Conservation Plan	Arizona HabiMap	Ironwood Forest National Monument Management Plan; City of Tucson Avra Valley Habitat Conservation Plan; Pima County Draft Multi-Species Conservation Plan, Arizona HabiMap.
3. Mitigates for all or most identified residual impacts that may warrant compensatory mitigation? √ for yes (1 point) or X for no (–2 points); Include justification.		√ Ecological Resources, Cultural Resources	√ Ecological Resources, Cultural Resources	√ Ecological Resources	√ Ecological Resources	√ Ecological Resources, Cultural Resources	√ Ecological Resources, Cultural Resources
4. Mitigation action (restoration/enhancement, acquisition, withdrawal, special designation, etc.).		Acquisition and/or active wildlife habitat restoration.	Acquisition, access management	Acquisition	Acquisition, restoration of floodplain function, erosion abatement, habitat restoration, and grazing control.	Acquisition	Acquisition and/or active wildlife habitat restoration
5. Site and its proposed actions meet regional conservation/mitigation goals and objectives? √ for yes (1 point) or X for no (–2 points)		√	√	√	√	√	√
Justification.		Protection falls within goals and objectives of the Pima County Sonoran Desert Conservation Plan.	Protection falls within goals and objectives of the Pima County Sonoran Desert Conservation Plan.	Conservation goals are consistent with the Pima County Sonoran Desert Conservation Plan.	Conservation goals are consistent with the Pima County Sonoran Desert Conservation Plan.	Conservation goals are consistent with the Pima County Sonoran Desert Conservation Plan.	Conservation goals are consistent with the Pima County Sonoran Desert Conservation Plan.
6. Proposed Mitigation Action and location Consistent with the Resource Management Plan? √ for yes (1 point) or X for no (–2 points); Include justification.		X Not in an acquisition area in Phoenix RMP or IFNM RMP.	X Not in an acquisition area in Phoenix RMP or in IFNM RMP.	X	X Not in an acquisition area in Phoenix RMP or IFNM RMP.	√ Acquisition in area identified in IFNM RMP.	X Not in an acquisition area in Phoenix RMP or IFNM RMP.

Regional Mitigation Strategy for the Arizona SEZs

Criteria		La Osa Ranch (Pima County)	Marana Mound (Pima County)	Ajo (Pima County)	Boa Sorte (Pima County)	Cocoraque Butte (Pima County)	Los Robles Archeological/Historic al District (Pima County)
	Candidate Sites						
7. Same HUC 4 watershed. Specify watershed.		X Middle Gila	X Middle Gila	√ Lower Gila	X Middle Gila	X Middle Gila	X Middle Gila
8. VRI Class and acres associated with each class.		No VRI	No VRI	No VRI	No VRI	No VRI	No VRI
9. Similar landscape value, ecological functionality, biological value, species, habitat types, and/or natural features. Score based on responses to criteria 9a and 9b. √ for yes (1 point) or X for no (−2 points).		√	√	√	√	√	√
9a. Current terrestrial landscape intactness score (use Sonoran Desert Rapid Ecological Assessment Data) and acres associated with each intactness category[47].		Very Low: 3,955 acres; Low: 428 acres; Mod Low: 856 acres; Mod High: 691 acres;	Very Low: 765 acres; Low: 1,736 acres; Mod Low: 1,798 acres; Mod High: 3,828 acres; High: 1,688 acres; Very High: 3,193 acres	Very Low: 247 acres; Low: 259 acres	Very Low: 497 acres; Mod Low: 979 acres; Mod High: 377 acres	Low: 564 acres; Mod Low: 351 acres; Mod High: 1,221 acres; High: 507 acres; Very High: 1,985 acres	Very Low: 1,839 acres; Low: 10 acres; Mod Low: 249 acres; Mod High: 426 acres; High: 365 acres; Very High: 247 acres
9b. Dominant vegetation communities.		Sonoran-Mojave Creosotebush-White Bursage Desert Scrub (36%) Sonoran Paloverde-Mixed Cacti Desert Scrub (34%)	Sonoran Paloverde-Mixed Cacti Desert Scrub (89%) Sonoran-Mojave Creosotebush-White Bursage Desert Scrub (7%)	Sonoran Paloverde-Mixed Cacti Desert Scrub (97%) Sonoran-Mojave Creosotebush-White Bursage Desert Scrub (2%)	Sonoran-Mojave Creosotebush-White Bursage Desert Scrub (58%) Sonoran Paloverde-Mixed Cacti Desert Scrub (32%)	Sonoran Mixed Paloverde Cacti Desert Scrub (99%) Sonoran-Mojave Creosotebush-White Bursage Desert Scrub (0.4%)	Sonoran Paloverde-Mixed Cacti Desert Scrub (47%) Agriculture (26%)
10. In SEZ Ecoregion √ for yes (1 point) or X for no (−2 points).		√	√	√	√	√	√
11. In SEZ ecological subregion. Specify subregion. √ for yes (1 point) or X for no (−2 points).		√ 1,000 acres of 81o, with minor additional 81K, 81L and 81n.	√ Mainly 81l, with some minor 81n.	√ 500 acres in 81k.	X 1158 acres of 81n, with 694 acres of 81l.	X 4340 acres of 81l, 235 acres of 81k, 53 acres of 81n.	X 2058 acres of 81o with 1066 acres and 81l and minor 81n.
12. Provides adequate geographic extent. Depending on whether site provides area for mitigation at least as large as the entire developable area of the SEZ. √ for yes (1 point) or X for no (−2 points).		√	√	X	X	√	√

[47] Intactness Categories: Very Low (−1.0 – −0.75); Low (−0.75 – −0.5); Mod Low (−0.5 – 0.0); Mod High (0.0 – 0.5); High (0.5 – 0.75); Very High (0.75 – 1.0).

Regional Mitigation Strategy for the Arizona SEZs

Criteria	Candidate Sites					
	La Osa Ranch (Pima County)	Marana Mound (Pima County)	Ajo (Pima County)	Boa Sorte (Pima County)	Cocoraque Butte (Pima County)	Los Robles Archeological/Historic al District (Pima County)
FEASIBILITY						
13. Feasibility of action	4	3	3	3	3	3
Justification of action feasibility. Stakeholder-provided score for 13a through 13e, as well as BLM expert knowledge, were used by the BLM to arrive at the overall Feasibility score, which is included in the overall site score.	Part of the property is being actively marketed for development, but there is a recognition that much of the site is not developable.	The land is predominantly owned by ASLD. Have to go to auction to purchase state lands. The private parcel owner has indicated a willingness to donate or sell this land to conserve the archaeological values.	Owners have indicated a willingness to sell. No existing site conditions that would preclude long-term conservation.	Owner has indicated a willingness to sell. Pima County has an appraisal. No existing site conditions would preclude long-term conservation.	Have to go to auction to purchase state lands.	The land is predominantly owned by ASLD. Have to go to auction to purchase state lands. There are no existing site conditions that would preclude long-term conservation.
13a. What level of documentation is available to demonstrate effectiveness of mitigation action? Use scale of 1 (little to no documentation) to 5 (well-documented).	4	5	3	3	4	3
Justification.	Research by Laura Jackson in Pinal County demonstrated feasibility of revegetating dry parts of the floodplain; NRCS has nearby Red Rock revegetation project which was successful; Ecological Opportunities in Lower Santa Cruz River, aerial photos show rapid revegetation of areas that were disturbed.	Marana Mounds is a well-researched archaeological site. The land is undeveloped.			Ironwood Forest National Monument Management Plan, Sonoran Desert Conservation Plan, City of Tucson Avra Valley Habitat Conservation Plan, Pima County Multi-species Conservation Plan	Ironwood Forest National Monument Management Plan, Sonoran Desert Conservation Plan, City of Tucson Avra Valley Habitat Conservation Plan, Pima County Multi-species Conservation Plan

Regional Mitigation Strategy for the Arizona SEZs

Criteria	La Osa Ranch (Pima County)	Marana Mound (Pima County)	Ajo (Pima County)	Boa Sorte (Pima County)	Cocoraque Butte (Pima County)	Los Robles Archeological/Historical District (Pima County)
			Candidate Sites			
13b.[48] Based on action required (e.g., restoration, BLM land management action, land acquisition, Congressional action), how difficult will implementation be? Use scale of 1 (difficult) to 5 (relatively easy).	4	4	4	4	4	4
Justification.	Part of this property is actively being marketed for development, but there is recognition that much of the site is not developable. Part of the area was previously graded. The Santa Cruz River here receives periodic natural flood flows and intermittent effluent discharges.	The land is predominantly owned by ASLD. The private parcel owner has indicated a willingness to donate or sell this land to conserve the archaeological values. Pima County would be willing to consider management of the property or holding a conservation easement on the property in perpetuity. There are no existing site conditions that would preclude long-term conservation.	The owners have indicated a willingness to sell. Pima County would be willing to consider management of the property or holding a conservation easement on the property in perpetuity. There are no existing site conditions that would preclude long-term conservation.	The owner has indicated a willingness to sell and Pima County has received an appraisal (and Phase 1 ESA report).		The land is predominantly owned by ASLD. There are no existing site conditions that would preclude long-term conservation.
13c. Time frame needed to establish site as mitigation location (estimated years).	2 years	2 years	2 years	2 years	2 years	2 years
13d. Time frame for achieving mitigation goals and objectives from implementation (estimated years).	5–10 years	2 years	0 years—achieving goals primarily with acquisition	5–10 years	0 years—achieving goals primarily with acquisition	0–10 years—depending on type of mitigation action
13e. Cost estimate.	Average cost per acre: $3,500 ($21 million – $42 million)	Average cost per acre: $3,500 (45.5 million)	$3,500 per acre ($1,750,000)	$3,500 per acre ($6.3 million)	$16,198,000	$3,500 per acre ($10,976,000)

[48] Rate the mitigation action, based on the following scale: restoration/enhancement actions (score of 5); BLM planning decisions (score of 3–4); land acquisition actions (score of 3–4); Congressional actions (score of 1). Ratings should be adjusted on the basis of factors such as cost of the action; time and effort requirements; public and/or BLM support for or opposition to action; and, for land acquisitions, willingness of seller.

Regional Mitigation Strategy for the Arizona SEZs

Criteria	Candidate Sites					
	La Osa Ranch (Pima County)	Marana Mound (Pima County)	Ajo (Pima County)	Boa Sorte (Pima County)	Cocoraque Butte (Pima County)	Los Robles Archeological/Historical District (Pima County)
EFFECTIVENESS / ADDITIONALITY						
14. Effectiveness and Additionality	4	4	3	3	3	3
Justification of effectiveness and additionality score. Scores for 14a through 14c were provided by stakeholders. BLM used these scores as well as their knowledge of the sites and actions.						
14a. [49] To what extent can the full spectrum of mitigation goals/objectives be met simultaneously? Use scale of 0 (low) to 5 (high).	5	5	4	3	5	5
Justification.	Preservation of military training, airspace and readiness; Waters of the US (Corps of Engineers); floodplain function and wetland mitigation; xeroriparian restoration; wildlife movement corridor protection; migratory bird mitigation; historic and archeological protection and interpretation; visual resource mitigation; recreational access; measures to abate wind and water erosion and manage invasive species.	Protection of this unique cultural and natural landscape would be an extremely effective means to mitigate solar impacts to other areas of importance to tribes, and is consistent with local plans.	Protection of cultural and natural resources	Restoration of floodplains, abatement of wind and water erosion of soils, revegetation.	Management of invasive plants; improvements for wildlife habitat; abatement of erosion; protection and interpretation of archeological and historical resources; conservation of a traditional cultural place.	Waters of the US (Corps of Engineers); floodplain function mitigation; mesquite bosque and wetland mitigation; other xeroriparian restoration; wildlife movement corridor; migratory bird mitigation; historic and archeological protection and interpretation; visual resource mitigation; recreational access; abatement of wind and water erosion.

---

49 Rate the extent to which the mitigation desired outcomes can be met simultaneously through mitigation actions at the site, based on the following scale: all (100%) of the goals and objectives can be met (score of 5); 75–99% can be met (score of 4); 50–74% (score of 3); 25–49% can be met (score of 2); less than 25% can be met (score of 1); none of the goals/objectives can be met (score of 0).

Regional Mitigation Strategy for the Arizona SEZs

Criteria	Candidate Sites					
	La Osa Ranch (Pima County)	Marana Mound (Pima County)	Ajo (Pima County)	Boa Sorte (Pima County)	Cocoraque Butte (Pima County)	Los Robles Archeological/Historical District (Pima County)
14b.[50] How effective will the mitigation be in the context of achieving mitigation goals/objectives for conserving/restoring ecosystem intactness? Use scale of 1 (low) to 5 (high).	5	5	5	3	5	5
Justification.	This site is not currently protected or managed for conservation.	This site is not currently protected or managed for conservation. In 2003, the Coalition for Sonoran Desert Protection included this candidate mitigation site in their proposal for federal protection.	This site is not currently protected or managed for conservation.	This site is not currently protected or managed for conservation.	This site is not currently protected or managed for conservation. Provides a full spectrum of mitigation opportunities to be met simultaneously.	This site is not currently protected or managed for conservation. Provides a full spectrum of mitigation opportunities to be met simultaneously.
14c. Mitigation consists of actions that would not otherwise be undertaken by BLM.	✓	✓	X	X	X	X
RISK						
15. Risk of action(s)	3	3	3	4	3	3
Justification of risk score. Scores for 15a through 15b were provided by stakeholders. BLM used these scores as well as their knowledge of the sites and actions.		State Land Dept. sale process requires auction	May face public resistance as a recreational use area	Relatively developed area	State Land Dept. sale process requires auction	State Land Dept. sale process requires auction
15a. What are the constraints or threats to success?	The land in question is part of the Santa Cruz River floodplain and portions of it could be encroached if not used for mitigation; the adjacent lands outside the floodplain are for sale and might be developable.	Without acquisition, ASLD could sell land for development, and recent residential development abuts the proposed Marana Mounds mitigation area.	Low, but while there is not an eminent threat of development in this area, these are some of the few, large tracts of private land available.	Land is in flood plain and not likely for development.	ASLD could allow leases which would damage some of natural and cultural values.	A portion of the land is part of the Santa Cruz River and Brawley Wash floodplain and could be encroached; the adjacent lands outside the floodplain will likely be developed.

[50] Rate the effectiveness of the mitigation actions at the site in terms of achieving mitigation goals/objectives, based on the following scale: highly effective (score of 5); moderately effective (scores of 2–4), and minimally effective (score of 1).

Regional Mitigation Strategy for the Arizona SEZs

Criteria	Candidate Sites					
	La Osa Ranch (Pima County)	Marana Mound (Pima County)	Ajo (Pima County)	Boa Sorte (Pima County)	Cocoraque Butte (Pima County)	Los Robles Archeological/Historical District (Pima County)
15b. What are surrounding land uses that will impact mitigation success (e.g., proximity to expanding urban areas, pressures on region for recreational land use, excessive groundwater withdrawal and drawdown conditions that could affect resources on the mitigation site)?	2 Residential development	2 The State Land identified here is adjacent to several Pima County natural open space properties, including Tortolita Mountain Park. Recent residential development abuts the proposed Marana Mounds mitigation area.	4 The surrounding land is primarily BLM-owned tracts.	4 Pima County Flood Control District is slowly buying land in the Brawley Wash floodplain. City of Tucson owns and manages much of the valley for protection of water resources.	4 Land is surrounded by Ironwood National Monument	3 Land outside of floodplain is likely to be developed at some point.
DURABILITY						
16. Durability of action(s).	3	3	3	3	5	3
Justification of durability score. Scores for 16a through 16b were provided by stakeholders. BLM used these scores as well as their knowledge of the sites and actions.	Not part of any existing ACEC or other area of protection as identified in the applicable RMP	Not part of any existing ACEC or other area of protection as identified in the applicable RMP	Not part of any existing ACEC or other area of protection as identified in the applicable RMP	Not part of any existing ACEC or other area of protection as identified in the applicable RMP	Monument designation makes this very durable	Not part of any existing ACEC or other area of protection as identified in the applicable RMP
16a. [51] How durable would the mitigation be from a time frame and management perspective? Use scale of 1 (low) to 5 (high).	5	5	5	5	5	5

---

[51] Rate the temporal and managerial durability of the mitigation action, based on the following scale: Congressionally protected lands would be very durable (score of 5); other federally administered lands specifically designated in land use plans or withdrawn by public land order would be moderately to very durable (score of 4–5); federally administered lands without any special designation but with enforcement oversight would have limited durability (score of 2); lands without special designation or enforcement oversight would not be very durable (score of 1).

Regional Mitigation Strategy for the Arizona SEZs

Criteria	Candidate Sites					
	La Osa Ranch (Pima County)	Marana Mound (Pima County)	Ajo (Pima County)	Boa Sorte (Pima County)	Cocoraque Butte (Pima County)	Los Robles Archeological/Historic al District (Pima County)
Justification.	The land could be acquired, added to the Monument with Presidential action, or administered through other BLM mechanisms such as an ACEC. A conservation easement could assist long-term protection.	The land could be acquired. A conservation easement could assist long-term protection. Access controls could include signage, management plan, rehabilitation of social trails.	The land could be acquired. A conservation easement could assist long-term protection.	The land could be acquired. A conservation easement could assist long-term protection.	Any lands acquired would become protected in the Monument without the need for amending the RMP.	The land could be acquired, could be added to the Monument, or administered through other BLM mechanisms such as an ACEC. A conservation easement could assist long-term protection.
16b. Are there potential effects of future climate change[52]?	Low	Low	Low	Low	Moderate	Low
PRELIMINARY SCORING Calculate score by summing the entries in blue-shaded cells. Scores are calculated based on entries in blue-shaded cells as follows: all scaled values (i.e., ratings from 1 to 5) are summed; 1 point is added for each √; 2 points are deleted for each X.	18	17	13	11	18	13
ADDITIONAL CONSIDERATIONS						
17. Presence of unique/valuable resources or features. (Up to 3 additional points for unique/valuable resources or features present at the candidate site, in 17a through 17h.)	2	2	2	2	2	2
17a. Perennial, protected sources of water.		No		Brawley Wash complex, with most of the area classified as an Important Riparian Area under the Conservation Lands System.	No	Santa Cruz River and Brawley Wash floodplain

[52] Climate change categories are from the Sonoran Desert Rapid Ecoregional Assessment future climate change model (BLM 2011).

Regional Mitigation Strategy for the Arizona SEZs

Criteria	Candidate Sites					
	La Osa Ranch (Pima County)	Marana Mound (Pima County)	Ajo (Pima County)	Boa Sorte (Pima County)	Cocoraque Butte (Pima County)	Los Robles Archeological/Historical District (Pima County)
17b. Unique species assemblages	Le Conte's thrasher, California leaf-nosed bat, and Sonoran Desert tortoise.	California leaf-nosed bat and Sonoran desert tortoise. The properties are in the historical distribution of Le Conte's thrasher.	Sonoran pronghorn and Sonoran desert tortoise observed on the properties.	California leaf-nosed bat. The properties are in the historical distribution of Le Conte's thrasher.	California leaf-nosed bat and Sonoran Desert tortoise; the site is also in the historical range of the Le Conte's thrasher.	California leaf-nosed bat and Sonoran Desert tortoise. The property is in the historical range for LeConte's thrasher.
17c. AZGFD Species of Greatest Conservation Need (distribution models)/Heritage Data Management System species (occurrence data)	52/-	42/-	41/-	44/-	38/-	48/-
17d. BLM categorized desert tortoise habitat	Not BLM land, no BLM category	Category 3 on edge	Category 2	Not BLM land, no BLM category	Category 2 and 3	Not BLM land, no BLM category
17e. T&E species or critical habitat and/or BLM sensitive species	T&E: Lesser long-nosed bat Critical Habitat: none BLM Sensitive Species: burrowing owl	T&E: Lesser long-nosed bat Critical Habitat: none BLM Sensitive Species: burrowing owl, cactus ferruginous pygmy owl	T&E: Lesser long-nosed bat, Sonoran pronghorn Critical Habitat: none	T&E: Lesser long-nosed bat Critical Habitat: none BLM Sensitive Species: burrowing owl, cactus ferruginous pygmy owl	T&E: Lesser long-nosed bat Critical Habitat: none BLM Sensitive Species: burrowing owl, cactus ferruginous pygmy owl	T&E: Lesser long-nosed bat Critical Habitat: none BLM Sensitive Species: burrowing owl, cactus ferruginous pygmy owl
17f. Desert washes (miles) or ephemeral playas (acres)	3,659 acres of additional non-developable floodplain along Santa Cruz River.	Not calculated, but does contain many acres xeroriparian habitat.	Gibson Arroyo	Brawley Wash complex, most of area classified as Important Riparian Area under the Conservation Lands System.	Riparian areas and intermittent water	Santa Cruz River and Brawley Wash floodplain

Regional Mitigation Strategy for the Arizona SEZs

Criteria	Candidate Sites					
	La Osa Ranch (Pima County)	Marana Mound (Pima County)	Ajo (Pima County)	Boa Sorte (Pima County)	Cocoraque Butte (Pima County)	Los Robles Archeological/Historical District (Pima County)
17g. Known highly significant and unique cultural resources.	41 acres of recorded archeological sites (where surveyed), including several large sites associated with the Santa Cruz River system. Juan Bautista de Anza National Historic Trail alignment passes through. The eastern ranchlands overlap with the Los Robles Archaeological District. Several related sites on nearby private lands were not originally included within the District because of landowner objections. All of these sites are considered ancestral sites by the Tohono O'odham.	The Tortolita Fan contains, by far, the most intact and best preserved prehispanic late Hohokam residential community, Marana Mounds, and its associated cultural landscape. This prehispanic cultural landscape is unique in that it is the only one of its kind that remains essentially intact and undeveloped.	The majority of the lands have not been surveyed for cultural resources, though due to the close proximity to the Tohono O'odham reservation, the properties likely have traditional cultural value to the Hia Ced O'odham community.	Proximity to the archaeologically rich Cañada Del Oro Wash.	Corcoraque Butte Archaeological District is listed on the National Register of Historic Places. Cocoraque Butte Archaeological District is adjacent to the Garcia Strip of the Tohono O'odham Indian Reservation, on the westernmost edge of Avra Valley. The Tohono O'odham Nation considers this butte to be a highly significant traditional cultural place with important spiritual values.	Juan Bautista de Anza National Historic Trail alignment passes through the site. The Los Robles Archaeological District) is listed on the National Register of Historic Places. Pima County's Sonoran Desert Conservation Plan identifies the Los Robles Complex as a Priority Archaeological Site Complex. All of these sites are considered ancestral places by the Tohono O'odham. The district encompasses 12,894 acres of State Trust lands and Bureau of Land Management lands in Pima and Pinal Counties.
17h. Other	373 acres of wildlife corridors; 2,122 acres of damaged land for mitigation and restoration; and scenic basalt-capped hills		High value as a wildlife corridor	Near the Bureau of Reclamation's Tucson Mitigation Corridor and along the axis of Brawley Wash.		Relatively unimpaired floodplain functions. Restoration of agricultural fields to proper floodplain function and wildlife habitat.
18. Links two or more protected areas √ for yes (1 point) or 0 for no (no score adjustment); include justification.	0 The floodplains and wildlife corridors connect to the Ironwood Forest National Monument.	√ Marana Mounds is adjacent to Pima County Tortolita Mountain Park on the east and BLM lands to the north.	0 Surrounded on 3 sides by BLM land. Close proximity to Cabeza Prieta NWR and Organ Pipe Cactus National Monument.	0	√ Acquires inholdings in Ironwood Forest National Monument.	√ Links City of Tucson conservation lands with Ironwood Forest National Monument.

Regional Mitigation Strategy for the Arizona SEZs

Criteria	Candidate Sites					
	La Osa Ranch (Pima County)	Marana Mound (Pima County)	Ajo (Pima County)	Boa Sorte (Pima County)	Cocoraque Butte (Pima County)	Los Robles Archeological/Historical District (Pima County)
COMBINED SCORE Add preliminary score to the additional consideration criteria in the blue-shaded cells. Scores are calculated based on entries in blue-shaded cells as follows: scaled values (i.e., ratings from 1 to 3) are summed; 1 point is added for each V.	20	20	15	13	21	16

This page intentionally left blank

www.ingramcontent.com/pod-product-compliance
Lightning Source LLC
Chambersburg PA
CBHW080654190526
45169CB00006B/2117